# Practical Hive

A Guide to Hadoop's
Data Warehouse System

Scott Shaw
Andreas François Vermeulen
Ankur Gupta
David Kjerrumgaard

***Practical Hive: A Guide to Hadoop's Data Warehouse System***

Scott Shaw
Saint Louis, Missouri, USA

Andreas François Vermeulen
West Kilbride North Ayrshire, United Kingdom

Ankur Gupta
Uxbridge, United Kingdom

David Kjerrumgaard
Henderson, Nevada, USA

ISBN-13 (pbk): 978-1-4842-0272-2
DOI 10.1007/978-1-4842-0271-5

ISBN-13 (electronic): 978-1-4842-0271-5

Library of Congress Control Number: 2016951940

Managing Director: Welmoed Spahr
Acquisitions Editor: Robert Hutchinson
Developmental Editor: Matt Moodie
Technical Reviewer: Ancil McBarnett, Chris Hillman
Editorial Board: Steve Anglin, Pramila Balen, Laura Berendson, Aaron Black, Louise Corrigan,
     Jonathan Gennick, Robert Hutchinson, Celestin Suresh John, Nikhil Karkal, James Markham,
     Susan McDermott, Matthew Moodie, Natalie Pao, Gwenan Spearing
Coordinating Editor: Rita Fernando
Copy Editor: Kezia Endsley
Compositor: SPi Global
Indexer: SPi Global
Cover Image: Designed by FreePik

Distributed to the book trade worldwide by Springer Science+Business Media New York, 233 Spring Street, 6th Floor, New York, NY 10013. Phone 1-800-SPRINGER, fax (201) 348-4505, e-mail orders-ny@springer-sbm.com, or visit www.springer.com. Apress Media, LLC is a California LLC and the sole member (owner) is Springer Science + Business Media Finance Inc (SSBM Finance Inc). SSBM Finance Inc is a Delaware corporation.

For information on translations, please e-mail rights@apress.com, or visit www.apress.com.

Apress and friends of ED books may be purchased in bulk for academic, corporate, or promotional use. eBook versions and licenses are also available for most titles. For more information, reference our Special Bulk Sales–eBook Licensing web page at www.apress.com/bulk-sales.

Any source code or other supplementary materials referenced by the author in this text is available to readers at www.apress.com. For detailed information about how to locate your book's source code, go to www.apress.com/source-code/.

Printed on acid-free paper

*I dedicate this book to my family. They put up with me being on the computer everyday and yet they have no idea what I do for a living. Love you!*

*—Scott Shaw*

*I dedicate this book to my family and wise mentors for their support. Special thanks to Denise and Laurence.*

*—Andreas François Vermeulen*

*I would like to express my gratitude to the many people who saw me through this book. Above all I want to thank my wife, Jasveen, and the rest of my family, who supported and encouraged me in spite of all the time it took me away from them.*

*—Ankur Gupta*

*"By perseverance, study, and eternal desire, any man can become great." —George S. Patton*

*—David Kjerrumgaard*

# Contents at a Glance

# Contents

# About the Authors

**Scott Shaw** has over 15 years of data management experience. He has worked as an Oracle and SQL Server DBA. He has worked as a consultant on Microsoft business intelligence projects utilizing Tabular and OLAP models and co-authored two T-SQL books by Apress. Scott also enjoys speaking across the country about distributed computing, Big Data concepts, business intelligence, Hive, and the value of Hadoop. Scott works as a Senior Solutions Engineer for Hortonworks and lives in Saint Louis with his wife and two kids.

**Andreas François Vermeulen** is Consulting Manager of Business Intelligence, Big Data, Data Science, and Computational Analytics at Sopra-Steria, doctoral researcher at University of Dundee and St. Andrews on future concepts in massive distributed computing, mechatronics, Big Data, business intelligence, and deep learning. He owns and incubates the "Rapid Information Factory" data processing framework. He is active in developing next-generation processing frameworks and mechatronics engineering with over 35 years of international experience in data processing, software development, and system architecture. Andre is a data scientist, doctoral trainer, corporate consultant, principal systems architect, and speaker/author/columnist on data science, distributed computing, Big Data, business intelligence, and deep learning. Andre received his Bachelor's degree at the North West University at Potchefstroom, his Master of Business Administration at the University of Manchester, Master of Business Intelligence and Data Science at University of Dundee, and Doctor of Philosophy at the University of Dundee and St. Andrews.

**Ankur Gupta** is a Senior Solutions Engineer at Hortonworks. He has over 14 years of experience in data management, working as a Data Architect and Oracle DBA. Before joining the world of Big Data, he worked as an Oracle Consultant for Investment Banks in the UK. He is a regular speaker on Big Data concepts, Hive, Hadoop, Oracle in various events, and is an author of the *Oracle Goldengate 11g Complete Cookbook*. Ankur has a Master's degree in Computer Science and International Business. He is a Hadoop Certified Administrator and Oracle Certified Professional and lives in London with his wife.

**David Kjerrumgaard** is a systems architect at Hortonworks. He has 20 years of experience in software development and is a Certified Developer for Apache Hadoop (CCDH). Kjerrumgaard is the author of *Data Governance with Apache Falcon and Cloudera Developer Training for Apache Hadoop*. He received his bachelor's and master's degrees in Computer Science from Kent State University.

# About the Technical Reviewers

**Ancil McBarnett** has been in the IT industry for over 20 years, where he initially began his "small data" career as an Oracle consultant and DBA in the Caribbean and Latin America. Ancil possesses an MBA with emphasis in Finance and a BSc. in Computer Science/Management.

Prior to working at Hortonworks he was the Architect Manager for a state agency responsible for sharing secure and sensitive data among first responder and justice systems and at Oracle, and was championing several Big Data and next generation Data Integration initiatives in a pre-sales capacity.

Since joining Hortonworks he has worked mainly with health providers who are looking to utilize Hadoop as the ideal platform to store and analyze secure data and to create modern data applications, with Hive as a pivotal tool to accomplish this.

You can find some of his articles on Hive and Tez tuning on the Hortonworks Community Connection.

**Chris Hillman** is Principal Data Scientist in the International Advanced Analytics team at Teradata. He has 20+ years of experience working in the business intelligence industry, mainly in the Retail and CPGN vertical, working as Solution Architect, Principal Consultant, and Technology Director. Chris works with the Teradata Aster Centre of Expertise and is involved in the pre-sale and start-up activities of Teradata Aster projects, helping customers understand whether MapReduce or SQL is an appropriate technique to use. Chris is currently studying part-time for a PhD in Data Science at the University of Dundee, applying Big Data analytics to the data produced from experimentation into the Human Proteome.

# Acknowledgments

Even before I joined Hortonworks I wanted to write a book on Hive. At the time there weren't many and the ones I saw where technically sound but not for the average users and especially not for someone coming from the relational database world. Once I began working at Hortonworks I figured it would be easy to sit down and write the book. I had all the best resources at my fingertips and access to some of the brightest people I've ever met. I had Hive committers like Alan Gates who never hesitated to answer an e-mail or spend a moment to talk to you at a conference. I had the friendship and support of the best Solution Engineering team in the world. Yet almost 2 and a half years later, there was still no book.

What I didn't predict was the incredible pace of this market and the herculean time commitment all of us on the team endure to provide solutions to our customers. It is truly a labor of love, but between work and family, the book had to wait. It waited a long time. I think any other publisher would have kicked me out the door and looked elsewhere, but Apress held steady (although I cannot honestly say they didn't push back a little and deservedly so) and trusted that someday we would have a book.

The struggle with writing a book on Hive is if you wait six months between writing then you're writing a new book. I came to terms that this was not the job of one person and I needed help. Ankur was one of the first to step up. If not for Ankur's perseverance and commitment, this book would not be in your hands right now. It was also Ankur who put us in touch with Andre and, I'm certain Ankur would agree, without Andre's incredible writing output and knowledge, you would also not have a book in your hands or, at the very least, it would be smaller and you would be much less informed. Finally, thank you to David, who has truly provided the technical exclamation point on the book and was vital to rounding out the edges and moving us forward.

There are countless other people who have helped in any way they could with little time they had. Cindy Gross from the Microsoft CAT team was an early participant and helped to keep the project moving forward. Thank you to Ancil for stepping up and helping with much needed technical reviews—especially on my chapters. But most especially thank you to Hortonworks for not only supporting the book but being downright excited about it. The greater Hortonworks team wasn't excited about the book just because it is a Hive book; they were excited for us, the team of authors, for our accomplishment. I never was forced to choose between my work and the book; it was my choice to focus on work.

Finally, thank you to my family. My kids may never have a need for Hive but I know they think it's pretty cool that dad help write a book. It's been a long journey from the days I was an English major to now being a Solutions Engineer for an open source Big Data company writing technical books, but I really do still look around me and count my blessings. I'll say it again I work with some of the brightest people in the industry and although I can't hold a candle to their intelligence, I do know their collective knowledge and insight makes me a better person.

—Scott Shaw

# Introduction

When I first learned about Hive I was working as a consultant on two data warehousing projects. One of them was in its sixth month of development. We had a team of 12 consultants and we were showing little progress. The source database was relational but, for some unknown reason, all the constraints such as primary and foreign key references had been turned off. For all intents and purposes, the source was non-relational and the team was struggling with moving the data into our highly structured data warehouse. We struggled with NULL values and building constraints as well as master data management issues and data quality. The goal at the end of the project was to have a data warehouse that would reproduce reports they already had.

The second project was smaller but involved hierarchical relationships. For example, a TV has a brand name, a SKU, a product code, and any number of other descriptive features. Some of these features are dynamic while others apply to one or more different products or brands. The hierarchy of features would be different from one brand to another. Again we were struggling with representing this business requirement in a relational data warehouse.

The first project represented the difficulty in moving from one schema to another. This problem had to be solved before anyone could ask any questions and, even then the questions had to be known ahead of time. The second project showed the difficulty in expressing business rules that did not fit into a rigid data structure. We found ourselves telling the customer to change their business rules to fit the structure.

When I first copied a file into HDFS and created a Hive table on top of the file, I was blown away by the simplicity of the solution yet by the far-reaching impact it would have on data analytics. Since that first simple beginning, I have seen data projects using Hive go from design to real analytic value built in weeks, which would take months with traditional approaches. Hive and the greater Hadoop ecosystem is truly a game-changer for data driven companies and for companies who need answers to critical business questions.

The purpose of this book is the hope that it will provide to you the same "ah-ha" moment I experienced. The purpose is to give you the foundation to explore and experience what Hive and Hadoop have to offer and to help you begin your journey into the technology that will drive innovation for the next decade or more. To survive in the technology field, you must constantly reinvent yourself. Technology is constantly travelling forward. Right now there is a train departing; welcome aboard.

# CHAPTER 1

■ ■ ■

# Setting the Stage for Hive: Hadoop

By now, any technical specialist with even a sliver of curiosity has heard the term Hadoop tossed around at the water cooler. The discussion likely ranges from, "Hadoop is a waste-of-time," to "This is big. This will solve all our current problems." You may also have heard your company director, manager, or even CIO ask the team to begin implementing this new Big Data thing and to somehow identify a problem it is meant to solve. One of the first responses I usually get from non-technical folks when mentioning Big Data is, "Oh, you mean like the NSA"? It is true that with Big Data comes big responsibility, but clearly, a lack of knowledge about the uses and benefits of Big Data can breed unnecessary FUD (fear, uncertainty, and doubt).

The fact you have this book in your hands shows you are interested in Hadoop. You may also know already how Hadoop allows you to store and process large quantities of data. We are guessing that you also realize that Hive is a powerful tool that allows familiar access to the data through SQL. As you may glean from its title, this book is about Apache Hive and how Hive is essential in gaining access to large data stores. With that in mind, it helps to understand why we are here. Why do we need Hive when we already have tools like T-SQL, PL/SQL, and any number of other analytical tools capable of retrieving data? Aren't there additional resource costs to adding more tools that demand new skills to an existing environment? The fact of the matter is, the nature of what we consider usable data is changing, and changing rapidly. This fast-paced change is forcing our hand and making us expand our toolsets beyond those we have relied on for the past 30 years. Ultimately, as we'll see in later chapters, we do need to change, but we also need to leverage the effort and skills we have already acquired.

Synonymous with Hadoop is the term *Big Data*. In our opinion, the term Big Data is slowly moving toward the fate of other terms like Decision Support System (DSS) or e-commerce. When people mention "Big Data" as a solution, they are usually viewing the problem from a marketing perspective, not from a tools or capability perspective. I recalled a meeting with a high-level executive who insisted we not use the term Big Data at all in our discussions. I agreed with him because I felt such a term dilutes the conversation by focusing on generic terminology instead of the truly transformative nature of the technology. But then again, the data really is getting big, and we have to start somewhere.

My point is that Hadoop, as we'll see, is a technology originally created to solve specific problems. It is evolving, faster than fruit flies in a jar, into a core technology that is changing the way companies think about their data—how they make use of and gain important insight into all of it—to solve specific business needs and gain a competitive advantage. Existing models and methodologies of handling data are being challenged. As it evolves and grows in acceptance, Hadoop is changing from a niche solution to something from which every enterprise can extract value. Think of it in the way other, now everyday technologies were created from specialized needs, such as those found in the military. Items we take for granted like duct tape and GPS were each developed first for specific military needs. Why did this happen? Innovation requires at least three ingredients: an immediate need, an identifiable problem, and money. The military is a huge,

**Electronic supplementary material**  The online version of this chapter (doi:10.1007/978-1-4842-0271-5_1) contains supplementary material, which is available to authorized users.

S. Shaw et al., *Practical Hive*, DOI 10.1007/978-1-4842-0271-5_1

complex organization that has the talent, the money, the resources, and the need to invent these everyday items. Obviously, products the military invents for its own use are not often the same as those that end up in your retail store. The products get modified, generalized, and refined for everyday use. As we dig deeper into Hadoop, watch for the same process of these unique and tightly focused inventions evolving to meet the broader needs of the enterprise.

If Hadoop and Big Data are anything, they are a journey. Few companies come out of the gate requesting a 1,000-node cluster and decide over happy hour to run critical processes on the platform. Enterprises go through a predictable journey that can take anywhere from months to years. As you read through this book, the expectation is that it will help begin your journey and help elucidate particular steps in the overall journey. This first chapter is an introduction into why this Hadoop world is different and where it all started. This first chapter gives you a foundation for the later discussions. You will understand the platform before the individual technology and you will also learn about why the open source model is so different and disruptive.

# An Elephant Is Born

In 2003 Google published an inconspicuous paper titled "The Google Filesystem" (http://static.googleusercontent.com/media/research.google.com/en/us/archive/gfs-sosp2003.pdf). Not many outside of Silicon Valley paid much attention to its publication or the message it was trying to convey. The message it told was directly applicable to a company like Google, whose primary business focused on indexing the Internet, which was not a common use case for most companies. The paper described a storage framework uniquely designed to handling the current future technological demands Google envisioned for its business. In the spirit of TL&DR, here are its most salient points:

- Failures are the norm

- Files are large

- Files are changed by appending, not by updating

- Closely coupled application and filesystem APIs

If you were a planning to become a multi-billion dollar Internet search company, many of these assumptions made sense. You would be primarily concerned with handling large files and executing long sequential reads and writes at the cost of low latency. You would also be interested in distributing your gigantic storage requirements across commodity hardware instead of building a vertical tower of expensive resources. Data ingestion was of primary concern and structuring (schematizing) this data on write would only delay the process. You also had at your disposal a team of world-class developers to architect the scalable, distributed, and highly available solution.

One company who took notice was Yahoo. They were experiencing similar scalability problems along Internet searching and were using an application called Nutch created by Doug Cutting and Mike Caffarella. The whitepaper provided Doug and Mike a framework for solving many problems inherent in the Nutch architecture, most importantly scalability and reliability. What needed to be accomplished next was a re-engineering of the solution based on the whitepaper designs.

---

■ **Note**    Keep in mind the original GFS (Google Filesystem) is not the same as what has become Hadoop. GFS was a framework while Hadoop become the translation of the framework put into action. GFS within Google remained proprietary, i.e., not open source.

---

When we think of Hadoop, we usually think of the storage portion that Google encapsulated in the GFS whitepaper. In fact, the other half of the equation and, arguably more important, was a paper Google published in 2004 titled "MapReduce: Simplified Data Processing on Large Clusters" (http://static. googleusercontent.com/media/research.google.com/en/us/archive/mapreduce-osdi04.pdf). The MapReduce paper married the storage of data on a large, distributed cluster with the processing of that same data in what is called an "embarrassingly parallel" method.

■ **Note** We'll discuss MapReduce (MR) throughout this book. MR plays both a significant role as well as an increasingly diminishing role in interactive SQL query processing.

Doug Cutting, as well as others at Yahoo, saw the value of GFS and MapReduce for their own use cases at Yahoo and so spun off a separate project from Nutch. Doug named the project after the name of his son's stuffed elephant, Hadoop. Despite the cute name, the project was serious business and Yahoo set to scale it out to handle the demands of its search engine as well as its advertising.

■ **Note** There is an ongoing joke in the Hadoop community that when you leave product naming to engineering and not marketing you get names like Hadoop, Pig, Hive, Storm, Zookeeper, and Kafka. I, for one, love the nuisance and silliness of what is at heart applications solving complex and real-world problems. As far as the fate of Hadoop the elephant, Doug still carries him around to speaking events.

Yahoo's internal Hadoop growth is atypical in size but typical of the pattern of many current implementations. In the case of Yahoo, the initial development was able to scale to only a few nodes but after a few years they were able to scale to hundreds. As clusters grow and scale and begin ingesting more and more corporate data, silos within the organization begin to break down and users begin seeing more value in the data. As these silos break down across functional areas, more data moves into the cluster. What begins with hopeful purpose soon becomes the heart and soul or, more appropriately, the storage and analytical engine of an entire organization. As one author mentions:

> By the time Yahoo spun out Hortonworks into a separate, Hadoop-focused software company in 2011, Yahoo's Hadoop infrastructure consisted of 42,000 nodes and hundreds of petabytes of storage (http://gigaom.com/2013/03/04/the-history-of-hadoop-from-4-nodes-to-the-future-of-data/).

# Hadoop Mechanics

Hadoop is a general term for two components: storage and processing. The storage component is the Hadoop Distributed File System (HDFS) and the processing is MapReduce.

■ **Note** The environment is changing as this is written. MapReduce has now become only one means of processing Hive on HDFS. MR is a traditional batch-orientated processing framework. New processing engines such as Tez are geared more toward near real-time query access. With the advent of YARN, HDFS is becoming more and more a multitenant environment allowing for many data access patterns such as batch, real-time, and interactive.

When we consider normal filesystems we think of operating systems like Windows or Linux. Those operating systems are installed on a single computer running essential applications. Now what would happen if we took 50 computers and networked them together? We still have 50 different operating systems and this doesn't do us much good if we want to run a single application that uses the compute power and resources of all of them.

For example, I am typing this on Microsoft Word, which can only be installed and run on a single operating system and a single computer. If I want to increase the operational performance of my Word application I have no choice but to add CPU and RAM to my computer. The problem is I am limited to the amount of RAM and CPU I can add. I would quickly hit a physical limitation for a single device.

HDFS, on the other hand, does something unique. You take 50 computers and install an OS on each of them. After networking them together you install HDFS on all them and declare one of the computers a master node and all the other computers worker nodes. This makes up your HDFS cluster. Now when you copy files to a directory, HDFS automatically stores parts of your file on multiple nodes in the cluster. HDFS becomes a virtual filesystem on top of the Linux filesystem. HDFS abstracts away the fact you're storing data on multiple nodes in a cluster. Figure 1-1 shows a high level view of how HDFS abstracts multiple systems away from the client.

Figure 1-1 is simplistic to say the least (we will elaborate on this in the section titled "Hadoop High Availability"). The salient point to take away is the ability to grow is now horizontal instead of vertical. Instead of adding CPU or RAM to a single device, you simply need to add a device, i.e., a node. Linear scalability allows you to quickly expand your capabilities based on your expanding resource needs. The perceptive reader will quickly counter that similar advantages are gained through virtualization. Let's take a look at the same figure through virtual goggles. Figure 1-2 shows this virtual architecture.

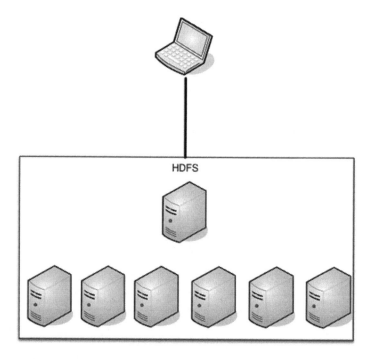

***Figure 1-1.*** *Simplistic view of HDFS*

Administrators install virtual management software on a server or, in most cases, a cluster of servers. The software pools resources such as CPU and memory so that it looks as if there is a single server with a large amount of resources. On top of the virtual OS layer we had guests and divide the available pool of resources to each guest. The benefits include maximization of IO resources, dynamic provisioning of resources, and high availability at the physical cluster layer. Some problems include a dependency on SAN storage, inability to scale horizontally, as well as limitations to vertical scaling and reliance on multiple OS installations. Most current data centers follow this pattern and virtualization has been the primary IT trend for the past decade.

■ **Note**   Figure 1-2 uses the term ESX. We certainly don't intend to pick on VMWare. We show the virtualization architecture only to demonstrate how Hadoop fundamentally changes the data center paradigm for unique modern data needs. Private cloud virtualization is a still a viable technology for many use cases and should be considered in conjunction with other architectures like appliances or public cloud.

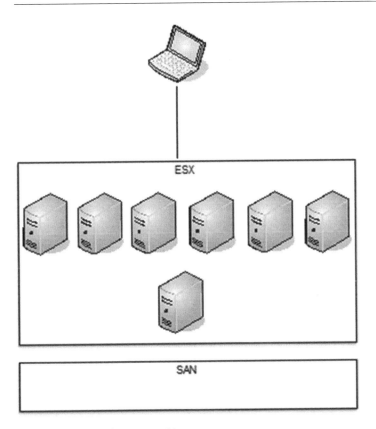

***Figure 1-2.*** *Virtualization architecture*

Other advantages include reduced power consumption and reduced physical server footprint and dynamic provisioning. Hadoop has the unenviable task of going against a decade-long trend in virtual architecture. Enterprises have for years been moving away from physical architecture and making significant headway in diminishing the amount of physical servers they support in their data center. If Hadoop only provided the ability to add another physical node when needed to expand a filesystem, we would not be writing this book and Hadoop would go the way of Pets.com. There's much more to the architecture to make it transformative to businesses and worth the investment in a physical architecture.

5

# Data Redundancy

Data at scale must also be highly available. Hadoop stores data efficiently and cheaply. There are mechanisms built into the Hadoop software architecture that allow us to use inexpensive hardware. As stated in the GFS whitepaper, the original design assumed nodes would fail. As clusters expand horizontally into the 100s, 1,000s, or even 10s of thousands, we are left with no option but to assume at least a few servers in the cluster will fail at any given time.

To have a few server failures jeopardize the health and integrity of the entire cluster would defeat any other benefits provided by HDFS, not to mention the Hadoop administrator turnover rate due to lack of sleep. Google and Yahoo engineers faced the daunting task of reducing cost while increasing uptime. The current HA solutions available were not capable of scaling out to their needs without burying the companies in hardware, software, and maintenance costs. Something had to change in order to meet their demands. Hadoop became the answer but first we need to look at why existing tools were not the solution.

## Traditional High Availability

When we normally think of redundancy, we think in terms of high availability (HA). HA is an architecture describing how often you have access to your environment. We normally measure HA in terms of nines. We might say our uptime is 99.999, or five nines. Table 1-1 shows the actual downtime expected based on the HA percentage (http://en.wikipedia.org/wiki/High_availability).

*Table 1-1.* *HA Percentage Summary*

| Availability Percent | Downtime Per Year | Downtime Per Month | Downtime Per Week |
|---|---|---|---|
| 90% ("one nine") | 36.5 days | 72 hours | 16.8 hours |
| 95% | 18.25 days | 36 hours | 8.4 hours |
| 97% | 10.96 days | 21.6 hours | 5.04 hours |
| 98% | 7.30 days | 14.4 hours | 3.36 hours |
| 99% ("two nines") | 3.65 days | 7.20 hours | 1.68 hours |
| 99.5% | 1.83 days | 3.60 hours | 50.4 minutes |
| 99.8% | 17.52 hours | 86.23 minutes | 20.16 minutes |
| 99.9% ("three nines") | 8.76 hours | 43.8 minutes | 10.1 minutes |
| 99.95% | 4.38 hours | 21.56 minutes | 5.04 minutes |
| 99.99% ("four nines") | 52.56 minutes | 4.32 minutes | 1.01 minutes |
| 99.995% | 26.28 minutes | 2.16 minutes | 30.24 seconds |
| 99.999% ("five nines") | 5.26 minutes | 25.9 seconds | 6.05 seconds |
| 99.9999% ("six nines") | 31.5 seconds | 2.59 seconds | 0.605 seconds |
| 99.99999% ("seven nines") | 3.15 seconds | 0.259 seconds | 0.0605 seconds |

Cost is traditionally a ratio of uptime. More uptime means higher cost. The majority of HA solutions center on hardware though a few solutions are also software dependent. Most involve the concept of a set of passive systems sitting in wait to be utilized if the primary system fails. Most cluster infrastructures fit this model. You may have a primary node and any number of secondary nodes containing replicated application binaries as well as the cluster specific software. Once the primary node fails, a secondary node takes over.

■ **Note**  You can optionally set up an active/active cluster in which both systems are used. Your cost is still high since you need to account for, from a resource perspective, the chance of the applications from both systems running on one server in the event of a failure.

Quick failover minimizes downtime and, if the application running is cluster-aware and can account for the drop in session, the end user may never realize the system has failed. Virtualization uses this model. The physical hosts are generally a cluster of three or more systems in which one system remains passive in order to take over in the event an active system fails. The virtual guests can move across systems without the client even realizing the OS has moved to a different server. This model can also help with maintenance such as applying updates, patches, or swapping out hardware. Administrators perform maintenance on the secondary system and then make the secondary the primary for maintenance on the original system. Private clouds use a similar framework and, in most cases, have an idle server in the cluster primarily used for replacing a failed cluster node. Figure 1-3 shows a typical cluster configuration.

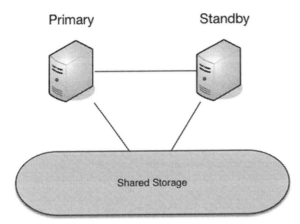

**Figure 1-3.** *Two-node cluster configuration with shared storage*

The cost for such a model can be high. Clusters require shared storage architecture, usually served by a SAN infrastructure. SANs can store a tremendous amount of data but they are expensive to build and maintain. SANs exist separate from the servers so data transmits across network interfaces. Furthermore, SANs intermix random IO with sequential IO, which means all IO becomes random. Finally, administrators configure most clusters to be active/passive. The passive standby server remains unused until a failure event. In this scenario hardware costs double without doubling your available resources.

Storage vendors use a number of means to maintain storage HA or storage redundancy. The most common is the use of RAID (Redundant Array of Independent Disks) configurations. Table 1-2 shows a quick overview of the most common RAID configurations.

**Table 1-2.** *The Most Common RAID Levels*

| RAID Level | Description | Fault Tolerance |
| --- | --- | --- |
| RAID 0 | Stripe array | None |
| RAID 1 | Mirror array | One disk |
| RAID 5 | Stripe with parity | One disk |
| RAID 1+0 | Striped mirrors | Multiple disks from one mirror |

RAID is popular due to the fact it provides data protection as well as performance enhancements for most workloads. RAID 0 for example supplies no data protection but speeds up write speed due to the increased amount of spindles. RAID, like clusters, come at a cost. In the case of mirrored RAID configuration you are setting aside a dedicated disk solely for the purpose of data recovery. Systems use the secondary disk only to replicate the data on write. This process slows down writes as well as doubling cost without doubling your storage capacity. To implement 5 TB of mirrored disk RAID, you would need to purchase 10 TB of storage. Most enterprises and hardware vendors do not implement RAID 0 or RAID 1 in server architectures.

Storage vendors such as EMC and NetApp configure their SAN environments with RAID 1+0 (RAID "ten"). This supplies the high-availability storage requirements as well as the performance capabilities. This works well for large SAN environments where arrays may consist of six or more drives and there may be dozens of arrays on the SAN. These arrays are carved up into LUNs (logical unit numbers) and presented to servers for use. These then become your mount points or your standard Windows drive letters.

---

■ **Note** Bear with me. The discussion around SANs and RAID storage may seem mundane and unimportant but understanding traditional storage design will help you understand the Hadoop storage structure. The use of SANs and RAID has been the de facto standard for the last 20 years and removing this prejudice is a major obstacle when provisioning Hadoop in data centers.

---

So, in essence SANs are large containers holding multiple disk arrays and managed by a central console. A company purchases a server, and then the server is provisioned in the data center with minimal storage (usually on a small DAS (direct attached storage) disk for the OS and connected via network links to the SAN infrastructure. Applications, whether point of sale applications or databases, request data from the SAN, which then pulls through the network for processing on the server. SANs become a monolithic storage infrastructure handing out data with little to no regard to the overarching IO processing. The added HA, licensing, and management components on SANs add significantly to the per-TB cost.

A lot of enhancements have been made in SAN technologies, such as faster network interconnects and memory cache, but despite all the advances the primary purpose of a SAN was never high performance. The cost per TB has dramatically dropped in the last 15 years and will continue to drop, but going out and buying a TB thumb drive is much different than purchasing a TB of SAN storage. Again, as with the virtualization example, SAN has real-world uses and is the foundation for most large enterprises. The point here is that companies need a faster, less expensive means to store and process data at scale while still maintaining stringent HA requirements.

# Hadoop High Availability

Hadoop provides an alternative framework to the traditional HA clusters or SAN-based architecture. It does this by first assuming failure and then building the mechanisms to account for failure into the source code. As a product Hadoop is highly available out of the box. An administrator does not have to install additional software or configure additional hardware components to make Hadoop highly available. An administrator can configure Hadoop to be more or less available, but high availability is the default. More importantly, Hadoop removes the cost to HA ratio. Hadoop is open source and HA is part of the code so, through the transitive property, there is no additional cost for implementing Hadoop as an HA solution.

So how does Hadoop provide HA at reduced cost? It primarily takes advantage of the fact that storage costs per terabyte have significantly dropped in the past 30 years. Much like a RAID configuration, Hadoop will duplicate data for the purpose of redundancy, by default three times the original size. This means 10 TB of data will equal 30 TB on HDFS. What this means is Hadoop takes a file, let us say a 1 TB web log file, and breaks it up into "blocks". Hadoop distributes these blocks across the cluster. In the case of the 1 TB log file, Hadoop will distribute the file using 24576 blocks (8192x3) if the block size is 128 MB. Figure 1-4 shows how a single file is broken and stored on a three-node cluster.

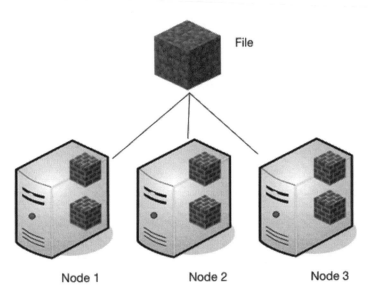

*Figure 1-4.  File broken into blocks, which are only portions of the total file*

Based on the configuration settings, these blocks can range between 128 MB and 256 MB!

---

■ **Note**    These are exceptionally large block sizes for a filesystem. As a reference point, the largest Windows block size, i.e. the largest size that can be read from disk into memory, is 4K. This is also the standard for most Linux-based OSs.

---

Large block sizes influence much of Hadoop's architecture. Large blocks sizes are core to how Hadoop is deployed, managed, and provisioned. Take into consideration the following factors influenced by large block sizes:

- Large files are more efficiently processed than smaller files
- There are fewer memory requirements on the master server (this will be discussed in the next section)
- Leads to more efficient sequential read and writes
- The seek rate is reduced as a percentage of transfer time

For the large file processing, let us go back to the 1 TB log file. Since the block size is set at 128 MB we get 24576 blocks sent over the network and written to the nodes. If the block size was 4K, the number of blocks would jump to 805306368 (268435456 x 3). As we will discuss later, this number of blocks would place undue memory pressure on specific portions of the cluster. The larger block size also optimizes the system for sequential reads and writes, which works best when considering dedicated drive access. A drive is simply a disk with a needle (aperture arm) moving across the surface (platter) to where the data is located. Storage makes no guarantee that data blocks will be stored next to each other on the platter so it takes time for the aperture arm to move randomly around the platter to get to the data. If the data is stored in large chunks or in sequential order, as is the case for most database transaction log files, then reading and writing becomes more efficient. The aperture only needs to move from point A to point B and not skip around searching for the data. Hadoop takes advantage of this sequential access by storing data as large blocks. When the time is spent by the aperture arm looking for data, this is called the seek rate. The two primary disk bottlenecks, and standard disks will always be the primary bottleneck, are seek rate and transfer time. The transfer time is the time it takes for the data to be moved from the disk into system memory. When compared to transfer time, seek rate is much slower. Hadoop reduces seek rate as a percentage of transfer time.

Storing large blocks may seem inefficient or restrictive on the surface, but Hadoop also has the concept of data locality to make the redundancy more useful. As mentioned earlier, Hadoop consists of a master node and worker nodes. We refer to the master node as the NameNode (NN) and we refer to the worker nodes as DataNodes (DN). The NameNode performs the following functions:

- Tracks which blocks in the cluster belong to which file

- Maintains where in the cluster each block is located

- Determines where to place blocks based on node location

- Tracks overall health of cluster through block reports

The NameNode not only breaks the file into blocks but it tracks where those blocks are placed in the cluster. Hadoop knows all the available DataNodes and on which rack the DataNodes are located. Knowing what rack nodes are on is called "rack awareness". Figure 1-5 takes the previous figure and expands it to include rack awareness.

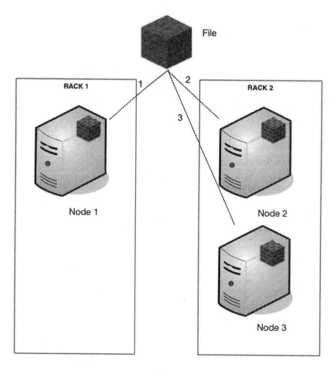

***Figure 1-5.*** *Writing blocks to HDFS with rack awareness*

Here are the steps Hadoop uses to write files:

1.  A block is written to node 1 in rack 1

2.  A copy of the block is written to node 2 in rack 2

3.  A copy is written to node 3 in rack 2

Even if there were more than two racks, the third block would still be written to the same rack as the second block. The order by which the blocks are written maximizes availability while reducing network traffic. By writing the second block to rack 2, HDFS immediately allows for the failure of an entire rack without affecting file recovery. The final write is to reduce network traffic because IO is much faster between nodes within a rack than between nodes in different racks. Files are large in HDFS so Hadoop has a number of different mechanisms to reduce network traffic. We will see more of this concept when we talk about processing.

Keep in mind that neither blocks nor files are stored on the NameNode. Data is only stored on the DataNodes. A client contacts the NameNode to determine where to write the blocks or where the blocks are located for read, after which the client talks directly to the DataNode. The NameNode stores the block information in memory. This is why large block sizes are important. The more blocks to track, the more memory the NameNode needs to store the information.

Only the NameNode knows where all blocks are located and to which file the blocks belong. If you lose a NameNode, you lose your cluster. This used to be a SPOF (single point of failure) factor for Hadoop, but now a NameNode can be effectively clustered for HA as you would with any other critical system. When building out your NameNode, you want to make sure the system has enough memory to handle the anticipated amount of blocks as well as having redundant hardware. DataNodes, on the other hand, do not need the additional hardware redundancy due to Hadoop's built-in redundancy. You will still want your DataNodes to have enough storage, memory, and CPU to hold and process the data.

# Processing with MapReduce

Storage is only part of the equation. Data is practically useless if we cannot process or analyze the data. Enterprises would be slow to adopt if they felt they were unable to derive insight from their mounds of stored data. We also do not want node failures negatively affecting our processing. Again, if we begin a job process on the cluster it would be unacceptable to have to restart the entire job five hours into processing simply because a single node became unavailable.

The first critical point to understand when discussing Hadoop processing is that Hadoop is a Java environment. Engineers who wrote Hadoop used the Java programming language. Hadoop processing, MapReduce, is also written in Java. In the early days of Hadoop, in order to do anything you had to have strong Java development skills. Luckily, for most of us, this is no longer the case. It can still be helpful to know and understand Java and how Java works both for writing MapReduce (MR) code as well as troubleshooting Hadoop, but as a business analyst or end user, you can now perform complex processing and analysis without ever having to touch Java code. As we will discuss further in the next chapter, engineers created Hive specifically to abstract away the necessity to write Java code.

If the market abstracts away Java and, by association, MapReduce, then why would it be necessary to understand how MapReduce processing works? The point is that the way MapReduce originally broke up large jobs into parallel executing tasks is still fundamental to distributed processing on a Hadoop cluster. Applications such as Hive, as well as an application like Pig, can still execute MapReduce behind the scenes (although it's not recommended) and it is helpful to understand what MapReduce is doing so that we can better tune our queries and understand their behavior. With the advent of YARN, MapReduce is just another means to access data on Hadoop, but MR is still important and worth discussing.

> ■ **Note** YARN stands for "Yet Another Resource Negotiator". YARN was developed by Arun Murthy at Hortonworks and is labeled as the "OS for Hadoop". It takes the resource administration away from the original MapReduce framework and allows MapReduce to focus on distributed processing instead of resource and task management. Cluster resource management is now generalized under YARN, which opens up other applications with different access patterns (interactive, real-time, as well as batch) to run simultaneously on the same cluster. YARN was introduced in Hadoop 2.x. Hadoop versions prior to 2.x are labeled as traditional Hadoop. Pre-YARN MapReduce is referred to as MRv1, while post-YARN MapReduce is referred as MRv2. YARN is discussed further in this chapter, but to dive deeper into YARN, we recommend reading *Apache Hadoop YARN* by Arun Murthy, Vinod Vavilapalli, Douglas Eadline, Joseph Niemiec, and Jeff Markham (Addison-Wesley, 2014).

As mentioned, Hadoop uses MapReduce specifically for processing data on a distributed network of computers. It does this by being what is called "embarrassingly parallel." This means the initial processing of the data occurs on separate nodes in parallel. This differs from traditional processing, which runs processing on a single computer or, in the case of database processing, pulls data from disk and stores it in memory for processing.

The Map phase is the first part of MapReduce parallel processing. Looking back on how Hadoop stores data on disk we remember it breaks a single file into multiple blocks. Each block contains a portion of the total data. So, if you have a 1 TB file with a list of names, that file will be broken into a large number of blocks with each block containing a subset of the names and these subsets stored on various nodes in the cluster. Figure 1-6 shows how a file containing names might be dispersed on a three-node cluster.

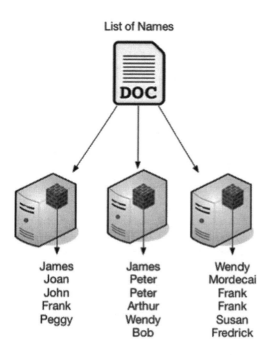

*Figure 1-6.* *List of company names distributed in blocks on a cluster*

Mapping in MapReduce is actually a Java function. It takes input and produces a new output. The output is a key/value pair.

---

■ **Note**    As you continue your journey into the Hadoop ecosystem, you will come across a lot of examples of the key/value pair concept. NoSQL focuses primarily on key/value structures. The reason this is important is because the key/value schema works great for distributed processing as well as processing semi-structured data not easily schematized into traditional RDBM systems.

---

In our example, a separate Map() function runs on each DataNode and processes all the blocks on that DataNode associated with the file. It does this independently of all the other blocks located on the other DataNodes. For the first node it will take the name James as input and output (James, 1). It will do this for each name in the block for each node so you would get the following output:

```
(James,1), (Joan,1), (John,1), (Frank,1), (Peggy,1)
(James,1), (Peter,1), (Peter,1),(Arthur,1),(Wendy,1),(Bob,1)
(Wendy,1),(Mordecai,1),(Frank,1),(Frank,1),(Susan,1),(Fredrick,1)
```

Keep in mind Hadoop processes each of these *in parallel*. There is no need for communication between nodes during the Map phase. This is critical when dealing with large data sets because you do not want inter-system communication or data transfer occurring between nodes. Introducing dependencies in processing can cause issues such as race conditions and deadlocks. By processing in parallel, Hadoop takes full advantage of dedicated IO resources in what is called *shared nothing* architecture.

Another key factor is the concept of taking the processing to the data. In our scenario, the Map task runs on the node where the data resides. The Map phase never pulls the data into a central location for processing. Again, this is key to processing large data sets since moving multi-terabytes or even petabytes amounts of data over the network would be impracticable. We want processing to occur on the nodes next to the data and utilize the full memory, disk, and CPU resources available to that node.

Once the Map phase completes, we have an intermediary phase called Shuffle and Sort. This phase takes all the key/value pairs from the Map phase and assigns them to a reducer. Each reducer receives all data associated with a single key. The Shuffle and Sort phase is the only time data is physically moved within the cluster and communication occurs between processes.

---

■ **Caution**    As we dig deeper into Hive performance we will want to focus on avoiding the reduce phase. This phase can be a bottleneck because it requires moving data over the network as well as communication between nodes. Also, the reduce phase cannot run until all mapping has completed.

---

Figure 1-7 shows how the data from the Map phase is moved across nodes by Shuffle and Sort.

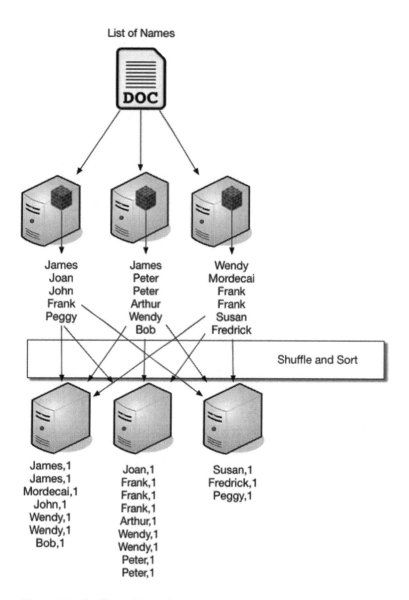

*Figure 1-7. Shuffle and Sort phase*

The Shuffle and Sort phase is responsible for sorting the data by keys and sending the data to the reducer tasks. Each reducer will receive all the data from a single key. For example, this means that one reducer will receive all the data from the name James. If there are 2 or 200 people named James, a single reduce job will still receive all the data associated with the key James. Notice the name Peter. The name occurs twice and each occurrence is on a single block of data. In the case of Peter, the data does not have to move to another node but can be mapped and reduced on the same node.

---

■ **Caution**    Know your data! If you have a data set with a disproportionate number of values for a single key, for example 50% of names in your file are Bob, then a single reducer may get overwhelmed.

---

The final stage is the Reduce phase. Reduce takes each key/value pair as input and produces a count aggregation based on the key. Those familiar with SQL can compare the reduce phase with a GROUP BY clause. The reducer will take (Frank,1,Frank,1,Frank,1) and convert it to (Frank,3). Figure 1-8 shows the final results.

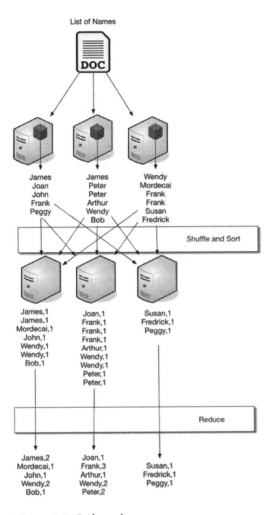

***Figure 1-8.*** *Reduce phase*

At the end of all our processing, we get a list of names and the total occurrence of each name in the file. This may seem trivial, but we can potentially run this MapReduce example on a 10 TB file with 100 or more nodes. As we add more nodes to the cluster our performance will improve. Traditional RDBMs just do not scale to this level.

# Beyond MapReduce

Mention has been made throughout this chapter on how MapReduce is not the only means to process data on Hadoop. MapReduce is an extraordinarily flexible parallel processing framework but, as scalable and flexible as it is, it also has many limitations. MapReduce processes data in batch. It exceeds at taking large data sets, processing them in parallel, and then aggregating the results. MapReduce does not work well with ad hoc or real-time query patterns. For example, if you want to get all sales for a product from every store in the past decade, and this query traverses 10 TB of data and you are willing to wait 10 hours for the results, MapReduce would be an excellent choice. But, if you want to get the top two items sold for five stores in Missouri and ten stores in Michigan, and you need that data in less than 10 seconds, then MapReduce is not a good solution. In reality, most organizations center around an ad hoc or near real-time processing business intelligence architecture of which MapReduce does not belong. Even simple SQL transactions using a small number of joins or GROUP BY clauses can take a long time to compute, especially when processing large amounts of data. We take for granted the speeds in which RDBMs process joins, GROUP BY, ORDER BY, and other computations and lose sight of the fact that the processing speeds are due to the upfront cost of constraining and conforming the data to specific schema structures and rules.

Hadoop is a *schema-on-read* as opposed to a *schema-on-write* framework. Ingesting data into traditional RDBMs involves transforming the data to fit into a relational structure comprising of tables, rows, and columns. Other structures also exist such as data types: int, varchar, date, as well as relational constraints between tables. The ETL (extract, transform, load) process works well, though still painful, when the source system is also relational. But what if your data is non-structured, or semi-structured? Log file data does not generally lend itself to a table structure. It is possible to transform the data into a relational model but at the cost of slowing down the data ingestion rate, as well as breaking the data ingestion process when simple domain constructs change, such as the adding columns or changing an integer value to a string. There is plenty of literature around the volume, velocity, and variety of modern data so I will not dive deeply into those ideas, but keep in mind that Hadoop is a framework conscious of the tradeoff between the traditional relational structure and the free-form process of ingesting data into a system that does not initially demand structure. Where we lose structure we gain flexibility. This is where Hadoop moves away from a simple storage environment and into a flexible and scalable compute environment that breaks down limitations between developers and rigid relational data structures.

Programmers write MapReduce tasks in Java. MapReduce handles the runtime complexities as well as the management and scheduling of jobs on the cluster. MapReduce requires a strong knowledge of Java and the MapReduce APIs. As Hadoop moves more mainstream, the product has had to move away from a Java development tool and cater more strongly to the areas of the business such as traditional ETL and business analytics which have dominated data analysis for the past 30 years. Adoption is key to the success of Hadoop and if everyone needs to learn Java to analyze data stored in Hadoop, overall adoption would be slow and difficult.

YARN has broadened the scope and flexibility of the Hadoop framework. YARN allows MapReduce to become only one method for accessing data stored on the Hadoop storage system. Other applications such as machine learning with Mahout and more recently Spark MLib, ad hoc querying with Hive and Tez, data flowing processing with Pig, and others can now execute side-by-side with MapReduce without any one application consuming all the cluster resources. YARN becomes fundamental to the adoption of Hadoop as an enterprise data store.

Your interest in this book indicates you may have a fundamental knowledge of the SQL query language. SQL is the language of traditional RDBMs and influences how we view and understand data access. All traditional relational database systems have a query engine whose purpose is to optimize access to structured data. Hadoop and MapReduce have limited knowledge of basic RDBMs constructs such as indexes, relational constraints, and statistics. Developers designed SQL query engines to take advantage of these assumptions and, if relational structures are not properly designed, do not exist, or are poorly implemented, performance significantly degrades. A larger question then becomes, "How do we match traditional RDBMs performance on a Hadoop cluster considering Hadoop is not architected like a traditional RDBMs"? This is the question being addressed by major Hadoop distributors as well as in the open community, and it is the one reason the community is moving away from batch-oriented MapReduce, and toward a more scalable and adaptable framework like YARN to allow for interactive and near real-time usage, in addition to batch.

## YARN and the Modern Data Architecture

So far we have discussed architectures around virtualization, SANs, traditional HA configurations, as well as disk configurations. These are fundamental concepts around data center design and standardization. Hadoop disrupts the notion of virtual servers, SAN storage, and RAID configuration. Vendors, data center administrators, as well as security administrators sometimes get nervous when asked to embark on this new way of storing and processing data. Let us also not forget the analysts who visualize and access the data for key business processes. The activities they perform are the gears moving the enterprise. They bring revenue and key insights to the business to drive new revenue channels and provide competitive advantage. Disrupting their activity means lost productivity and lost revenue.

A disruptive technology such as Hadoop inevitably stirs up backlash and FUD in many camps. Vendors will fight, and rightfully so, to maintain their data center footprint and argue for the advantages of their technology and the disadvantages of others. Other vendors try to embrace the inevitable implementation while assuming a key role in the play. While the storm of feature/function and risk/reward rages in the trenches, CIOs, CTOs, as well as business analysts just want the data efficiently and cheap as well as with minimal disruption.

The primary job of the Hadoop community and the vendors in the space (we will discuss vendors and distributions in more detail in the next chapter) is making minimal disruption a reality. Vendors, salespeople, and solution engineers can easily get mired in the feature debate and lose sight of the reason why Hadoop was created. Hadoop, at its essence, is a platform or architecture driving modern analytics. Industry refers to this as the *Modern Data Architecture*.

Figure 1-9 shows components of the Modern Data Architecture.

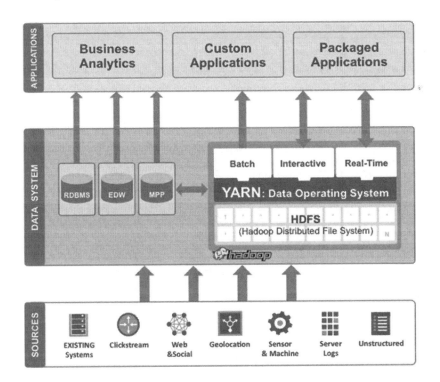

***Figure 1-9.*** *Hadoop as part of an existing data architecture solution*

The architecture incorporates additional sources into the data flow that were previously untapped due to the restrictions of traditional RDBMs. We can now include sources such as clickstream, web and social, sensor and machine, logs, and images. As we pull this data into Hadoop as streaming inputs or batch, we stage them in HDFS for direct analysis or movement into other systems. This approach optimizes the RDBMs, EDW, and MPP resources by offloading resource intensive and time-consuming extract, transform, and load operations onto the much more economical Hadoop platform. You essentially move from an ETL model to a ELT model. You extract and load everything into Hadoop but only transform the data appropriate to your given platform or analytical needs.

YARN is the driving force behind this architecture. As previously mentioned, prior to the introduction of YARN the only computation engine for Hadoop was MapReduce. MapReduce has many benefits but also many restrictions. Traditional Hadoop places MR jobs into a queue and a job cannot run until the previous job finished. This was due to the notion of slots and how many of these slots were available for an MR job to run. MR jobs were batch operations taking hours or days to finish. MR is great if you use your cluster for answering singular Big Data questions, but if you wanted to analyze daily sales at the same time you are drilling through a dashboard, you were out of luck.

YARN introduces the idea of containers. Containers are a pool of resources such as CPU, storage, and memory dedicated to a specific application process. A ResourceManager schedules jobs and arbitrates application resources based on assigned policies. These policies may or may not include such things as "Marketing gets a maximum of 50% of the cluster memory" or "50% of the cluster memory is assigned to marketing and HR and of that HR gets 30%". These key constraints allow for cluster resource provisioning on a user or group basis.

---

■ **Note** The example given in the text would be an example of the Capacity Scheduler. This scheduler allows for the granular allocation of resource on a per group or user level. Another example scheduler would be the Fair Scheduler, which behaves as a FIFO (first-in-first-out) scheduler or, more simply, an equal opportunity scheduler. The default scheduler for YARN is the Capacity Scheduler.

---

DataNodes run an ApplicationMaster whose purpose is to control each container on a per-application basis. The ApplicationMaster acts as the messenger for the ResourceManager, more specifically a component of the RM called the ApplicationManager, and controls resource allocation locally on each node. This allows the YARN framework to scale better than if the ResourceManager were a central manager of all node resources and did not have the benefit of local resource negotiators.

The ApplicationMaster adds a benefit in that third-party products can write applications that utilize the AM design and their application will run in conjunction with other AM applications. As seen in Figure 1-9, the introduction of the YARN framework and the AM daemon allows for multi-use query access such as batch, interactive, and real-time processing. We refer to this as multi-tenancy and it is the foundation of the Modern Data Architecture and why enterprises can now begin building a data lake to stage their data for whichever analytical tool they choose to swim in its waters. Integration is key to companies adopting Hadoop and to the implementation of the Modern Data Architecture. The original spirit of Hadoop and of YARN drives this integration in that the development of both is open and available for the benefit of all.

## Hadoop and the Open Source Community

We cannot discuss Hadoop, YARN, or Hive without mentioning open source software development as well as how open source software fits into enterprise adoption. Open source has always been a key component of Hadoop and the Hadoop ecosystem. When we say ecosystem, we are referring to all the applications that integrate directly with Hadoop and are part of the Apache Software Foundation (ASF). This includes Hive but also includes other features such as Sqoop, Pig, Oozie, Flume, and dozens of others. Each one represents a distinct software development project within ASF. This distinction is important because it can cause confusion when determining compatible versions as well which features are available for which product versions. Luckily for us all project development for each product is executed in the open and we are able to freely follow conversations around feature enhancements as well as bug fixes. Besides the lack of software licensing, this is what makes open source software truly "open". Development is not hidden away due to proprietary nondisclosure, in fact anyone can add to the discussion or recommend features that should be included in future product releases.

A number of large software companies like Microsoft have contributed open source code. Companies that have large Hadoop installations have also contributed code back into the product. What incentivizes them to make code public? The driver behind open source software development is the idea that by contributing code back to the project, the product innovates faster and everyone benefits from the innovations of the community. In addition, becoming a committer to an open source project is not a bad thing to have on your resume.

As you begin your Hive journey, you will be spending a majority of time on the ASF home page for Hive. This page can be found at `http://hive.apache.org`. Figure 1-10 shows the home page.

### GENERAL
Home
Downloads
License
Privacy Policy

### DOCUMENTATION
Language Manual
Javadoc
Wiki

### COMMUNITY
Becoming a Committer
Edit Website
How to Contribute
Resources for contributors
Issue Tracking
Mailing Lists
People

### DEVELOPMENT
Builds
Design Docs
FAQ
Hive JIRA
Precommit Patch Testing
Version Control

### PMC
ByLaws
How to Release

### ASF
Donations
Sponsorship
Thanks
Website

## APACHE HIVE TM

The Apache Hive ™ data warehouse software facilitates querying and managing large datasets residing in distributed storage. Hive provides a mechanism to project structure onto this data and query the data using a SQL-like language called HiveQL. At the same time this language also allows traditional map/reduce programmers to plug in their custom mappers and reducers when it is inconvenient or inefficient to express this logic in HiveQL.

### Getting Started

Check out the Getting Started Guide on the Hive wiki.

### Getting Involved

Hive is an open source volunteer project under the Apache Software Foundation. Previously it was a subproject of Apache Hadoop, but has now graduated to become a top-level project of its own. We encourage you to learn about the project and contribute your expertise. Here are some starter links:

- Give us feedback: What can we do better?
- Join the mailing list: Meet the community.
- Become a Hive Fan on Facebook.

***Figure 1-10.*** *ASF home page for Hive*

Key links are the Language Manual and Wiki under Documentation as well as Hive JIRA under Development. JIRA is issue-tracking software developed by Altassian and is used by the ASF community to track bugs, issues, and general project management cases associated with a product. Look at JIRA as the helpdesk for ASF software projects. We will talk more about Hive ASF details in later chapters, but it is first important to get a good understanding of the open source process and what it means to projects such as Hadoop.

The following quote underlines the purpose of ASF:

> *"The Apache Software Foundation provides support for the Apache community of open source software projects, which provides software products for the public good...The Apache projects are characterized by a collaborative, consensus based development process, an open and pragmatic software license, and a desire to create high quality software that leads the way in its field."*

ASF is an organization supporting various software development projects. It does this by providing to the community a repository and development methodology, as well as forums and support channels used by the community to create applications in the open. It provides for a central location for the community of programmers to monitor and regulate software development. The emphasis is on a "collaborative consensus," which means decisions are voted on by individuals who, in turn, were voted on to have the ability to control the process. Their position and power within a project is a direct result of their contributions and leadership.

Each Apache project is independent from one another and each project will have top-level PMCs (Project Management Committee) assigned to it and who control the overall project direction. An individual can be part of a PMC in multiple projects but it is rare and discouraged. Directly below a PMC is a Committer, who has write access to the project. Committers are essentially the project developers and they submit code to the project. Here is the list of Hive committers: `http://people.apache.org/committers-by-project.html#hive`. A Committer can also be a release manager, which is someone who is responsible for the logistics behind major releases. At the lowest level is a Contributor. A Contributor is someone who may ask a relevant question or make a good suggestion. Contributors have no authority in project direction and they are unable to add or make changes to code.

This certainly is not to say that contributors are not important. This is a volunteer-based organization, although Committers are highly sought after and organizations are more than willing to pay top salaries to have them on their payroll. Projects still need contributors who are willing to put in personal time to help with everything from documentation and bug reports, to basic evangelism. You do not have to be a seasoned developer or live in Silicon Valley to be a contributor. Contributors, as well as Committers, come from all walks of life and reside all over the world. Keep in mind that open source development is a community. It is a community of dedicated, driven volunteers who enjoy creating world-class software for the benefit of everyone and, if a company decides to pay you a high salary because your development skills have been proven and accepted by a community of developers in the open source meritocracy, so much the better. Also keep in mind that if you are a frequent contributor and you contribute valuable code then you can be voted on to become a full-fledged committer.

Every decision made on a project is made in the mailing list. Nothing is secret and it can be extremely fascinating, albeit time-consuming, to follow these conversations. You can find the Hive mailing lists here: `http://hive.apache.org/mailing_lists.html`. Hive has four separate lists: User, Developer, Commits, and Security. User is a general list for questions and support and is monitored by developers but is primarily a user-to-user forum. I strongly suggest subscribing to this mailing list if you plan to use Hive (which I assume you will). Simply click on the `user-subscribe@hive.apache.org` link and send an empty e-mail. You should receive e-mail verification of your subscription.

---

■ **Caution**    Subscribing to e-mail lists is helpful and informative but can generate a lot of "noise". The Hive community is vibrant and active and you will be able to see a lot of support and use case activity in the listings that you will not find anywhere else on the Internet. You can always unsubscribe if you find the information unhelpful or overwhelming. Another option to get help on Hive is the Hortonworks Community Connection or HCC. You find it by going to `http://community.hortonworks.com`.

---

The developer, commits, and security lists may be too esoteric for those who care more about Hive as an analytic platform than Hive as a development project. For the purpose of understanding concepts in this book and using Hive on a day-to-day basis, there is no need to subscribe to those lists, although feel free if you want to see the inner workings. You also have access to the e-mail archives you can view without subscribing. It is easy to get deep in the weeds when following projects, especially when discussions move to bug fixes or code development.

The Apache Software Foundation provides governing policies around code development. These policies are democratic in nature, though a bit stricter in the majority wins type of democracy (think Congressional policy). Committers and the community vote code commits as well as package releases and procedural policies. In many cases only PMC members have binding votes. Votes, like development decisions, are performed online via public forums. If you agree with the commit you type a +1, you do not agree you type a -1, which essentially acts as a veto. Besides the standard -1, 0-, +1 votes, the following list shows fraction votes and their meaning.

- +0: I don't feel strongly about it, but I'm okay with this.

- -0: I won't get in the way, but I'd rather we didn't do this.

- -0.5: I don't like this idea, but I can't find any rational justification for my feelings.

- ++1: Wow! I like this! Let's *do* it!

- -0.9: I *really* don't like this, but I'm not going to stand in the way if everyone else wants to go ahead with it

- +0.9: This is a cool idea and I like it, but I don't have time/the skills necessary to help out.

A vote of -1 kills the process until the veto is resubmitted as an approval or withdrawn. The individual casting the veto must also submit a technical design document explaining the reason for the veto. This helps cut down the chance of people abusing the veto policy. The veto option provides the process with a strong system of checks and balances whereby a single person has the ability to have their dissents and arguments fully addressed in an open forum. Only once all parties agree does the PMC change or release the code.

These rules differ depending on whether or not the vote is on code change, procedural policy, or a new release. We will not go into detail as to these difference, just know that the process is based on a democratic foundation designed to produce the best software possible. The process allows everyone the chance to contribute opinions and ideas to the project while building consensus and agreement to the project direction and functionality. An individual's status in a project is based on meritocracy. Peers elect a committer or PMC based on their contributions and demonstrated knowledge around the product. The open source community is truly a community of the best and brightest whose primary purpose is to develop better software for everyone.

## Where Are We Now

I will not deny the Hadoop landscape is changing faster than any single book can follow. Release cycles are measured in months and not years. Patches and updates are measured in weeks, not months. The open source community innovates faster than anything we have ever witnessed before. Adoption drives innovation. As large companies, maybe like yours, take on the challenge and opportunity of Hadoop and all it has to offer, they find defects or must have items. These same companies, actually the hard working developers and engineers in those companies, maybe like you, have the opportunity to drive the innovation by committing code, submitting JIRAs, or offering suggestions through your Hadoop vendor to smooth out edges and further drive innovation and adoption. The open source community is vibrant, innovative, driven, and committed to providing high quality, but mostly ingenious, software solutions solving complex modern data problems.

A small but critical component of this ecosystem is Hive. Hive is critical because it is the entry point into an exceedingly complex data storage environment. Hive is the link between the traditional and the new. Hive is the nod by the Hadoop development community that 40 years of RDBM design and access is of value and useful and worth the effort in order to drive adoption.

---

■ **Note**    I focus on 40 years because E.F. Codd first published his paper "A Relational Model of Data for Large Shared Data Banks" in June of 1970. Oddly, but probably not coincidentally, that whitepaper was published out of San Jose, which is the same area as the original Google GFS paper that influenced Hadoop development at Yahoo, located near the same area.

---

Hive is Hadoop access for the masses. Hadoop for the masses is no more negative or less pragmatic than the advent of the Ford Model T or the microwave. I personally hope the trend continues and I think it will. Hadoop for all its scalability and redundancy is nothing without adoption by the users who actually perform analysis and insight in an organization. Data is nothing if it is not useful, easily accessible, or provides immediate ROI. SQL is the natural language for data and the obvious choice for general Hadoop analysis. SQL provides ease of use, common understanding, and flexibility. Hive, though not 100% mapped to ANSI SQL, takes core parts of traditional SQL and allows business analysts to quickly adapt to and function on the Hadoop environment.

Other SQL on Hadoop engines exist such as Impala, HAWQ, and Spark SQL. Each has its benefits and drawbacks, areas of strength and areas of weakness. All of them, including Hive, understand the value of providing interactive SQL capabilities on Hadoop along with the performance we expect from traditional business intelligence infrastructures. Hive stands out with its widespread adoption and diverse development community represented by some of the largest IT organizations in the world. Hive is not going away and, as we will see in more detail in the coming chapters, continues to grow in features and capabilities with the singular purpose of empowering business user to unlock insight stored in Hadoop.

# CHAPTER 2

■ ■ ■

# Introducing Hive

As much as the Hadoop ecosystem evolves and provides exceptional means to access new types of data and structures, we cannot deny the influence and purpose of traditional relational systems. Relational systems and especially the data access methods employed by these systems have served as a valuable tool for over 30 years. The SQL query language brought data access to the masses by abstracting away concepts such as data location and instead allowed developers to focus on how the data will be presented. SQL excels as a declarative language in which you clearly specify what you want to do in simple English language syntax. You SELECT, JOIN, SUM, data FROM a source WHERE the value equals, or does not equal, something. The developer does not have to worry about where the data resides on disk, and the structure of the data is already predefined in a relational format consisting of tables with rows and columns.

The attraction of SQL to the Hadoop world was not in its ability to consume data schematized as rows and columns or its efficient use of indexes and statistics but instead, SQL's popularity as a data query tool. Simply put—a lot of people who accessed data knew how to write SQL statements. Keep in mind that early Hadoop adoption involved HDFS as the storage system and MapReduce as the compute framework. Java is the language of MapReduce so early in the Hadoop adoption, if you needed to perform computation and access data in Hadoop, you had to write Java code, specifically MapReduce programs. Large companies like Facebook came to realize that you could not hire enough Java developers to write the amount of MapReduce code needed to take full advantage of the quantity of data stored in HDFS. In order to increase adoption and ease of use, developers needed to abstract away MapReduce complexity in favor of a more demotic programming language.

The answer was SQL (Structured Query Language). MapReduce, originally, would stay as the compute language but would be relegated to behind-the-scenes functionality. Hive or, more precisely, HiveQL became a language a business analyst could adopt because the syntax looked similar to SQL, yet it could take advantage of the embarrassingly parallel processing power of MapReduce. Interactive SQL on Hadoop became the concept behind Hive and the language itself is called HiveQL.

---

■ **Note** I have not been able to find any history behind why Hive was decided as the name of the project. The original whitepaper (`http://www.vldb.org/pvldb/2/vldb09-938.pdf`) has no mention of the reasoning behind the name. In addition, Facebook adopted Hive as an abstraction layer for MapReduce which, at the time, was the only compute option for HDFS. Other engines have since been introduced that are more interactive than MR, but SQL is still the most widely used abstraction layer.

---

© Scott Shaw, Andreas François Vermeulen, Ankur Gupta, David Kjerrumgaard 2016
S. Shaw et al., *Practical Hive*, DOI 10.1007/978-1-4842-0271-5_2

Facebook acknowledged limitations in the initial design of HiveQL. Originally, Hive was an abstraction layer and not a panacea for the inherent limitations of MapReduce as a batch-orientated compute architecture. What we will see in subsequent chapters is the evolution of Hive and HiveQL as a framework capable of running on more traditional, i.e. familiar, interactive query engines. The recent evolution of Hive has moved it away from simply an abstraction layer running on top of batch-centric MapReduce to a framework capable of utilizing the full functionality of what we have come to expect from an interactive query engine. As we will see in later chapters, Hive has developed from a simple SQL veneer over MapReduce to a fully functional interactive framework running on a performant query engine as well as cost-based optimizers and file level statistics.

The chapter provides a quick overview of Hadoop distributions with the primary intent to standardize on a specific offering we will use in the book. It is easy to get bogged down in the various offerings and it would distract from the main topic of Hive if we were to continually show each example running in each distribution. Just be aware that besides any discussion around the Tez engine, most of the code provided can be executed in any distribution. In addition, the architecture of Hive and of clusters in general is universal and applicable across the board. Though briefly covered in this chapter, the topics of Hive architecture are discussed in more detail in Chapter 3.

# Hadoop Distributions

Before diving into Hive's architecture, we first need to address the proverbial "elephant" in the room around Hadoop in general. Hive's open source roots, as well as Hadoop and other ASF projects, poses some complexity when considering your install and configuration options. There are a number of different approaches and we cannot cover all of them in this book. Well, we could but then the book would not be much fun and we would take that much longer to actually begin using Hive.

We can break down Hive deployment options into two basic categories: roll-your-own or use a distribution. The roll-your-own option is a term used to mean downloading your own binaries and installing all the components yourself. The open source nature of the products allows you to download the full products as you see fit without any regard to a traditional user license. This means you will not need to pay a fee or even give away any personal information and, most importantly, a salesperson will not call you. The tradeoff to this approach is the complexity and the need for increased administration skills, especially around Linux and general Linux software build procedures. But also, and most importantly, having to deal with the interoperability between the release of Hive you download and other applications.

---

▨ **Caution**   If you are not familiar with the world of open source, you will quickly realize that as much as the open source community excels in passion and innovation it just as much lacks in standard documentation. The quality of documentation can vary wildly from one project to another. In some projects, key concepts and steps are omitted due to the misconception of the audience's technical level and background. Overall though I think open source documentation has gotten much better and some projects have better documentation than even proprietary offerings.

---

If you are a business end user wanting to try out the product or run tutorials, I do not recommend this approach. You will spend far too much time mucking around with Linux administration problems. Plus, documentation is limited at best and sometimes completely non-existent. Versioning can also be intimidating. Project development is in isolation from one product to another so the most recent version of each project is not necessarily compatible with one another. As of this writing, Table 2-1 shows the versions for the various ASF projects used in the three most recognized Hadoop distributors: Cloudera, MapR, and Hortonworks.

**Table 2-1.** *Apache Project Versions*

| Project Name | Cloudera CDH 5.7 | MapR 5.1 | Hortonworks 2.4 | Current Release |
|---|---|---|---|---|
| Accumulo | N/A | N/A | 1.7.0 | 1.7.1 |
| Atlas | N/A | N/A | 0.5.0 | 0.5.0 |
| Ambari | N/A | N/A | 2.2.2 | 2.2.2 |
| Calcite | N/A | N/A | 1.2.0 | 1.7.0 |
| Crunch | 0.11.0 | N/A | N/A | 0.14.0 |
| DataFu | 1.1.0 | N/A | 1.3.0 | 1.3.0 |
| Falcon | N/A | N/A | 0.6.1 | 0.6.1 |
| Flume | 1.6.0 | 1.6.0 | 1.5.2 | 1.6.0 |
| Hadoop | 2.6.0 | 2.7.0 | 2.7.1 | 2.7.2 |
| Hbase | 1.2 | 1.1 | 1.1.2 | 1.2.1 |
| Hive | 1.1.0 | 1.2.1 | 1.2.1 | 2.0.1 |
| Impala | 2.5.0 | 2.2.0 | N/A | 2.5.0 |
| Knox | N/A | N/A | 0.9.0 | 0.9.0 |
| Mahout | 0.9.0 | 0.11.0 | 0.9.0 | 0.12.1 |
| Oozie | 4.0.0 | 4.2.0 | 4.2.0 | 4.2.0 |
| Phoenix | 4.3.0 | N/A | 4.4.0 | 4.7.0 |
| Pig | 0.12.0 | 0.15.0 | 0.15.0 | 0.16.0 |
| Ranger | N/A | N/A | 0.5.0 | 0.6.0 |
| Sentry | 1.5.1 | N/A | N/A | 1.6.0 |
| Slider | N/A | N/A | 0.80.0 | 0.90.2 |
| Solr | 5.2.1 | 4.10.3 | 5.2.1 | 6.0.1 |
| Spark | 1.6.0 | 1.6.1 | 1.6.0 | 1.6.1 |
| Sqoop | 1.4.6 | 1.4.6 | 1.4.6 | 1.4.6 |
| Storm | N/A | 0.9.4 | 0.10.0 | 1.0.1 |
| Tez | N/A | N/A | 0.7.0 | 0.8.3 |
| Zookeeper | 3.4.5 | 3.4.5 | 3.4.6 | 3.4.8 |

ASF projects are not required to perform QA compatibility between versions; this is the job of the Hadoop distributors. The three primary vendors are Hortonworks, MapR, and Cloudera. Each vendor supplies an easy startup "sandbox" platform you can use to get up and running quickly with Hadoop and the ecosystem.

■ **Notes**    Table 2-1 is not an exhaustive listing of all features available in each distribution. I have highlighted only the features that exist on ASF as either top-level or incubator projects and are considered standard features in one or the other distribution. Projects listed as N/A does not mean the distribution does not have that functionality. It primarily means the functionality is handled by a non-ASF solution. Projects are constantly being added and updated. My guess is in the time it takes this book to be published and reach your hands, the versions will have already changed.

Each distribution will add functionality dependent on how well the community adopts the new feature and where Cloudera and MapR choose to provide proprietary solutions. Spark, for example, when it first was released, came standard in CDH, but Hortonworks provided it as a technical preview, until recently. Spark is not a standard offering across all distributions. MapR uses a proprietary version of Apache Hadoop called MapFS. Both Cloudera and MapR include Hive and but tend to focus on their own creations, called Impala and Drill respectively.

Distributions are a work in progress and always evolving. They are organic and grow as technology features mature and wane. They focus on ease of setup and integration as well as operations, governance, and security. In the end, distributions provide a solid technical standard and come with world-class engineering support to help you along your Hadoop journey.

# Cluster Architecture

Before we jump specifically into Hive, we first need to quickly address cluster design and set some performance expectations as well as general practices as you build and grow your Hadoop cluster. This is a Hive book and not a Hadoop architecture book, so we will take a high-level view of how to design a cluster. We will also review some key terminology used in Hadoop clusters. This will hopefully help you better navigate and understand the platform Hive depends on to operate.

Because data volumes always increase and there are always use cases in the pipeline, your cluster will grow. It may grow slowly over a period of months, or it may grow rapidly as your company brings on new use cases and lines of business (LOB) become excited about the possibility of new analytics and insight. Setting things correctly from the onset will help prepare you for unexpected growth. Luckily, Hadoop was designed to grow, i.e., to scale easily to meet your needs.

Architecting your cluster involves determining on which nodes to place which components. Where you install services is critical because it affects both cluster availability as well as cluster performance. Generally, administrators divide cluster servers into three categories: master, edge, and worker. A master server contains any component considered absolutely critical to the health of the cluster and usually involve components where high availability is a requirement. A worker server contains any cluster service that is easily replaced or can incur downtime without fear of data loss. Following are examples of services you will want to provision on master node(s) in a typical cluster.

- NameNode
- JobTracker
- ResourceManager
- Secondary NameNode
- HBase Master, HiveServer2
- Oozie Server

- Zookeeper

- Storm Server

- WebHCat Server

Hadoop vendors tend to segregate clusters into three sizes: small, medium, and large. How many nodes constitute a small, medium, or large is mostly a heuristic exercise. Some say if you can manage a 50-node cluster then you can manage a 1,000-node cluster. Generally speaking, a small cluster will tend to be less than 32 nodes, a medium cluster is between 32 and 150 nodes, and a large cluster is anything over 150. Another common design template is whether your cluster will fit on a single rack or multiple racks. A small cluster fits on a single rack while medium to large clusters will span multiple racks.

Again, these are generalizations and your mileage may vary but it is likely that if you have more than 32 nodes or multiple racks in your cluster, you are dealing with many more Hadoop components and interfaces than in a smaller cluster and, in addition, your company has decided that Hadoop will be a core platform in your organization and with critical functionality. These additional components will require additional resources as well as more focus on disaster recovery, high availability, and security across the stack. You will also have to pay close attention to network configurations. These include the speed of your top-of-rack switches as well as the bandwidth between inner-rack nodes.

Small clusters will have more components running on a single server than larger clusters. As your cluster grows, you will want to think about segregating master components and providing dedicated nodes to them. Small clusters are ideal for proof-of-concepts, pilots, or development environment. Think about using a cloud service for these types of clusters since they can be extremely affordable and quickly implemented. Cloud providers such as Google, Microsoft's Azure, and Amazon's AWS all have quick and easy methods for standing up small and large clusters. You can choose to run your cluster with minimum administration as a PaaS (Platform as a Service) or, if you want more control, as a IaaS (Infrastructure as a Service).

A hardware discussion is out of scope for this book and, in any case, any mention of hardware specifications would only be quickly outdated. Hadoop has matured enough and garnered enough interest in data centers that all the hardware vendors provide reference architectures for Hadoop clusters. Many of the hardware and chip vendors have chosen to partner with each of the vendors.

---

■ **Note**   Hadoop doesn't require major vendor hardware. Feel free to go to the janitor closest or your local resale shop and grab the cheapest boxes you can find, although getting permission to install these in your data center may be a bit more difficult. Keep in mind that as much Hadoop touts its resiliency, master servers are still SPOF (single point of failure) and need to be accounted for appropriately.

---

The Hive client is installed on all worker nodes. When interacting with Hive, you will most likely access it through a web portal such as Ambari or Hue. These servers tend to be installed on edge nodes. Edge nodes have fewer resources with no master server components. Keep in mind that they may contain metadata repositories that should be backed up like any other relational database system. You can think of edge nodes as management servers or even web servers. An edge node may contain the operational software such as Ambari, MCS, or Cloudera Manager as well as client components such as Pig or Hive. They may also be used for firewall purposes such as is the case for Apache Knox. The point being is edge nodes tend to be smaller servers whose main purpose is to act as a client gateway into the larger Hadoop infrastructure. You may still want to provision edge nodes with a fair amount of storage due to the amount of potential application logging that can occur.

Another way to look at edge nodes are as management servers that contain non-distributed components. For example, Ambari runs as a single instance and is not distributed across multiple nodes. The NameNode has the same feature. Because these components are vital to the cluster but not distributed, a management node will need to be designed with fault tolerance in mind. Management servers also tend to be much more RAM sensitive than storage. You do not normally need much storage for a management server. Figure 2-1 shows a simple diagram of a client, a management node, and worker nodes and the components traditionally stored on each.

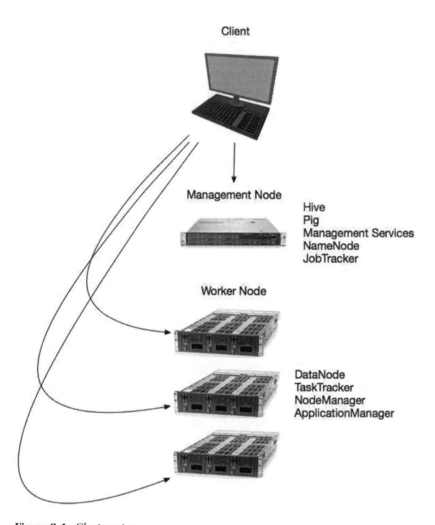

*Figure 2-1.* *Cluster setup*

The lines in Figure 2-1 indicate ways in which a client can interact with the cluster. You may access the management tools directly on the management node through a RESTful API or web browser. The other method would involve transferring data to HDFS by contacting the NameNode for file and block information and then working directly with the DataNodes. Again, there are many factors that would make your overall cluster design much more complicated and increase the amount of servers you may deploy. For example, you may want a dedicated HiveServer2 instance or a server used as a Knox security gateway. You may even have a federated cluster with multiple NameNodes or an additional NameNode for high-availability failover. These are all good discussions to have between your internal operations and security teams and your distribution vendor.

The point being is that Hive is only one small component of a larger Hadoop ecosystem. No one builds out a Hadoop cluster just to run Hive. Clusters are built for innumerable reasons from ETL offload to ingesting and persisting streaming sensor data. Clusters will most likely include applications like Solr, which is used for text searching, or HBase, which is used for more transaction-like processing. Cluster design and tuning deserves a book on its own but know that Hadoop cluster is a versatile platform environment meant to change the way your organization manages, stores, and analyzes all your data.

# Hive Installation

Despite all the features and functionalities packed into a Hadoop distribution, our focus is on Hive. As of the writing of this book, Hive 2.0.1 is the latest Apache version. Although the latest Apache version is 2.0.1 we will work exclusively with version 1.2.1 of Hive throughout subsequent chapters because it is the latest version tested and offered in a distribution. If you happen to already be using CDH 5.7, which uses Hive 1.1 with patches, most of the functionality should still work. Functionality involving the Tez engine will not be available because Cloudera does not support Tez as a SQL engine.

Throughout this book we will focus on the Apache version of Hive found in the Hortonworks distribution. The reason for this approach is because the open source version of Hive is the de facto gold standard version. Any addition or subtraction from the core bits does not fully represent what the open source community meant Hive to be or, more specifically, contains proprietary additions and/or omissions unique to a given distribution. Hortonworks is the distribution most closely aligned to the Apache version. Cloudera provides Hive in their distribution but their SQL-on-Hadoop solution is primarily focused on Impala. MapR standardizes on Apache Drill based on Google's Dremel. It is beyond the scope of this book to go into a feature comparison between the products. Just know you have many choices but not all the solutions are necessarily mutually exclusive. For our purposes, we will stick to Hive and to the Apache version to ensure all the features and options discussed in this book perform as expected.

There are a number of ways to "enter" Hadoop. Few readers of this book will have access to a multi-node cluster. If you are one of these people, you may want to skip ahead to the next chapter, or you can follow along and install your own personal environment. There is certainly something to be gained in going through even a basic setup and install process. For the rest of you, we assume you will perform at least one of these options:

- Install Hadoop and Hive using Apache code from Apache.org.

- Install Hadoop and Hive using the documentation instructions on a vendor's site such as Hortonworks, Cloudera, or MapR.

- Install Hadoop and Hive using the virtual sandbox from MapR, Cloudera, or Hortonworks.

- Install Hadoop in a cloud offering such as Google, Azure, or AWS.

Out of these four options, I strongly recommend the third or the fourth options. The sandbox will be the option we will focus on in this book. I do not want to de-emphasize the ease of using a Cloud offering. Each cloud provider has distributions in their marketplace, which makes setting up a cluster a trivial exercise. Most cloud providers also have automated sandbox installs. If you happen to have an account with a cloud provider, we strongly recommend using that environment throughout this book. If you do decide to install Hive manually outside a distribution, the overall install is trivial compared to installing the full Hadoop application, although Hive still requires a cluster for any data processing. You have the choice to install Hive either through a GZIP file or through a project builder such as Maven that builds Hive from the source code. The apache.org Hive web site has all the steps to build your Hive environment at `https://cwiki.apache.org/confluence/display/Hive/GettingStarted`.

---

■ **Note**    We will bring this up again because we know certain readers will feel something is lacking in the text, but this is a Hive book and not a "how to install Hadoop" book. Most books currently on the market have a section on how to install Hadoop from source but, in many cases, those instructions are incomplete or out-of-date once the book is published. This book focuses on the easiest method to get started so that you can quickly get your Hive environment up and running.

---

Installing a distribution's virtual environment will work fine for the purposes of this book. You will not need a fully functional and highly available cluster to run through the exercises presented here. We will not be overly concerned with performance. You will want to have enough storage and processing in order to run the VM and store the necessary data sets. The typical virtual Hadoop sandbox environment has the following out-of-the-box requirements:

- Virtual machine application: VMWare or VirtualBox

- Minimum 8 GB of RAM

- At least 1 GB of storage

- Minimum 2 vCPU

As always, more is better but we are trying to test functionality, not performance. In the real world, you would not want to try to crunch TB or PB of data in Hive on a single node cluster. You may need to think about increasing available RAM if you choose to play with other features such as Hbase, which require more processing. We have tried to make the data sets used in this book large enough to be interesting but still small enough to work practically on an average workstation. You have the ability to customize more data set sizes so, if you choose, feel free to work with larger data sets for testing and additional insight. You may stress your workstation compute resources, but you will not stress Hive and Hadoop.

A distribution download can be as much as 8.5 GB. As mentioned, the primary distribution used in this book will be the Hortonworks sandbox. The Hortonworks sandbox does not require software licenses. No software licensing allows for a much better testing and development experience for anyone just starting out with the technology because you are not restricted to only using the product within a given time period or you do not have access to all the tools. The authors of this book do not intend to dissuade you from downloading and working with the other distributions in order to get a good feeling for both the similarities as well as the differences. Each distribution will have Hive but Hortonworks is a major backer of the Hive initiative and invests most heavily in Hive development.

Hive is a client application using HDFS for its backend storage. Included in Hive are other server and functional components such as HiveServer2 and HCatalog. The details around these and other structures will be discussed in Chapter 3, "Hive Architecture". For now, just know that installing Hive is essentially installing a client application on your Hadoop cluster. You will need to designate nodes for the Hive client as well as the Hive metastore (HCatalog) and HiveServer. Each one will run as a separate service. Figure 2-2 shows what the services look like through the Ambari 2.2.2 console. When running the Hortonworks sandbox you can connect to Ambari via your local loopback address or by the sandbox.hortonworks.com DNS address plus the Ambari port number. Type the following in a browser, preferably in Firefox or Chrome: http://sandbox.hortonworks.com:8080.

**Figure 2-2.** *Hive services in Ambari*

Note that five Hive-related services are running: Hive Metastore, HiveServer2, MySQL Server, WebHCat Server, and Hive Client. All of these services are necessary for Hive to operate and each one will be discussed in more detail in later chapters.

Each service in the summary is installed on a single node. The sandbox runs Hadoop in what is referred to as *pseudo-distributed mode*. This essentially fools the Hadoop system into thinking it is running on a cluster when, in fact, it is running on only a single node. Hadoop replication is set to one (default is 3 on a multi-node cluster), which means for our installment we are not concerned with fault-tolerance or high availability. This works fine for the purposes of our demonstrations and examples.

Whether you choose Ambari, MCS, or CM, each of these products provide a means to manage the Hive services as well as alter and view configuration settings. Within each you can stop and start services, view running queries and jobs, and check the resource health of the nodes. Each is an operational application used not only to manage Hive but also any other service component running on the cluster. Since you are mostly likely the sole owner of your personal Hadoop installation running on your desktop, you will need to be familiar with administrating the environment. You will mostly be using the operation tools to alter Hive's running configuration files. As a developer or business analyst at your business though, you will not have much reason to work with these tools. It is still advantageous to be familiar with what options they provide to maintain the health of your environment and get the most out of the product.

# Finding Your Way Around

Now that you have a green light, or green icon, on all your Hive services, you are ready to use Hive as an SQL-on-Hadoop tool. Ambari views provide an easy means for executing Hive queries through a graphical user interface. You can use another third-party application like SQuirrl SQL (see http://squirrel-sql.sourceforge.net/) that connects to your Hive metastore. In our exercises we will be using HiveQL through the command-line (CLI) as well as the Ambari Hive view environment. These are development tools that allow you to execute SQL queries against Hive tables as well as import custom UDFs or SerDes. They are not analytic tools! There is a broad range of analytics that can connect to Hive through ODBC or JDBC connections. Some of the more popular ones will be discussed in a later chapter.

■ **Note**    For now do not worry about terms like SerDe or UDF. Some of you already familiar with SQL will understand a bit about what a user defined function (UDF) is used for and they are not that much different in Hive. SerDe is a different concept that we will talk about and use in later chapters.

As mentioned, there are two primary ways of interfacing with Hive: command-line and Ambari views. Figure 2-3 shows the Ambari Hive view in Ambari 2.2.2.

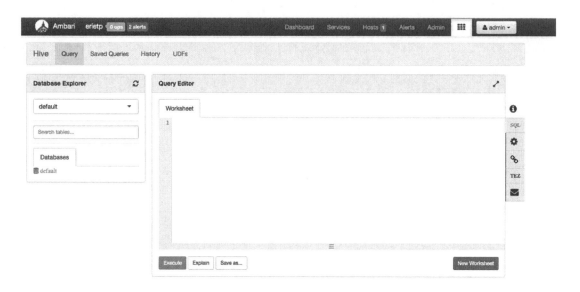

***Figure 2-3.***  *Ambari view*

The interface should be intuitive to anyone familiar with SQL query tools. The main parts include the ribbon, a database explorer, query editor, and various configuration and management options. You will type HiveQL statements into the query editor and then press the Execute button to run the queries. You will use the toolbar to view saved queries as well as see a history of executed queries. Keep in mind that Hive has hundreds of configuration settings. You have the option of changing environment settings at runtime or managing the configurations through the Hive service in Ambari. Some of these settings will be addressed in later chapters. The Ambari Hive view is designed for an end user and not an administrator. Typically a business analyst or SQL developer will use the Hive view for executing and testing queries against their data sets.

The database explorer window functions similarly to the USE command in SQL. You will be able to select any database you have access to and each database contains its own list of tables. If no database is specified, then Hive uses a database called *default*. To see what tables are in the database, you can either select the database or execute a *show tables* query in the query editor.

If you have installed a sandbox you should see two tables, called sample_07 and sample_08. While attached to the default database, execute the following query in the query editor. After typing your query, press the Execute button.

```
SELECT * FROM sample_07;
```

Hive SQL is called HiveQL. When you are executing from the Hive view query editor, the semicolon at the end of the statement is optional. As we will see later when executing HiveQL from the command-line, Hive requires the semicolon at the end of each statement. HiveQL also does not recognize upper and lowercase characters. For the purposes of readability, we will show all HiveQL-specific commands in uppercase. Figure 2-4 shows the query output.

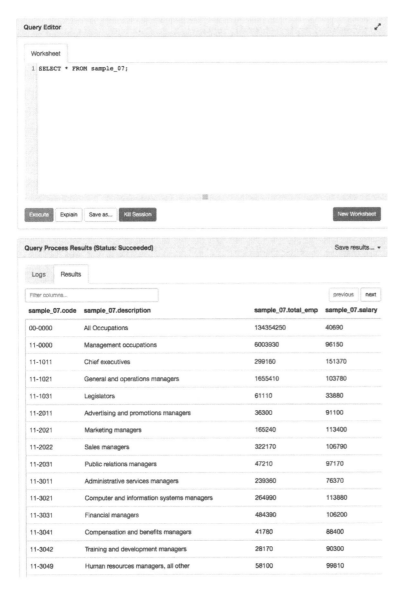

**Figure 2-4.** *Results from the sample_07 query*

As you create tables and databases in Hive, they will appear in HCatalog. HCatalog provides a means for other applications besides Hive to access these tables, preventing you from having to recreate a table on a per-application basis. A HCat table, as well as any Hive table you create, can be accessed via an ODBC or JDBC connection as well as through specific HCat loaders. The details around HCat and connecting to Hive tables will be discussed in later chapters. Just know for now that HCat tables and Hive tables are essentially one in the same and Hive is the means to create schemas for data files stored in Hadoop.

The magic of Hive is not in how it differs from traditional SQL but in how it is similar. Based on the query syntax as well as the results, you would not have a clue you may be running the command against a raw data file broken into blocks across 100s of nodes in a distributed architecture. Additionally, you may be querying across terabytes of data with response times similar to what you would see with gigabytes of data on a traditional relational system.

Of course, Hive out-of-the-box will not be as performant as an RDBMs. It is equivalent of running a query with no indexes. There are a number a performance best practices you will want to be familiar with, such as using ORC files, Hive indexes, and table partitions. These will be discussed in the chapter on Hive performance. Remember too that Hive is an analytic tool and will not replace your existing OLTP processes. This goes as much for the processing expectations as well as the similarities with the ANSI SQL and HiveQL. For example, you can expect to see windowing functions but you will not see triggers. As scalable as Hive is, it does not mean you can start using Hive as an e-commerce cart application, at least not yet.

# Hive CLI

Besides the graphical interface option, Hive provides a command-line interface for managing and running scripts, data-definition commands, and data-manipulation commands. The command line provides flexibility and low overhead for interacting with Hive.

Hive CLI is great for quick-and-dirty SQL work or easy scripting. This section will not go into deep detail about Hive CLI but it will show you have to get started. To connect to Hive CLI, you will need to ssh into the sandbox using the following command:

```
ssh root@sandbox.hortonworks.com -p 2222
```

At the command prompt type your password. This will start a ssh session on the sandbox and you will be logged in as the root user. At the command line, type hive. After some initial configurations displays, your command prompt should now show as >hive. Here are the step-by-step commands to log into the sandbox and start HiveCL.

```
HW10882:~ sshaw$ ssh root@sandbox.hortonworks.com -p 2222
root@sandbox.hortonworks.com's password:
Last login: Sun Jun 12 17:14:05 2016 from 10.0.2.15
[root@sandbox ~]# hive
WARNING: Use "yarn jar" to launch YARN applications.

Logging initialized using configuration in file:/etc/hive/2.4.0.0-169/0/hive-log4j.
properties
hive>
```

You can execute all your normal HiveQL commands from the prompt. The one difference is that you will be required to end all your statements with a semicolon. Pressing Enter will start a new line if a semicolon is not present. If you happen to press Enter without a semicolon, you can go ahead and type a semicolon on the new line and Hive will execute the statements on the previous lines.

Let's start by typing show tables; on the command line and pressing Enter. The following code demonstrates the show tables command as well as a select from the sample_07 table. Notice we put a LIMIT command on the query. This acts just like SQL and restricts the amount of rows to the value set by the LIMIT command.

```
hive> show tables;
OK
sample_07
sample_08
Time taken: 7.651 seconds, Fetched: 2 row(s)
hive> SELECT * FROM sample_07 LIMIT 10;
OK
00-0000   All Occupations                   134354250      40690
11-0000   Management occupations            6003930        96150
11-1011   Chief executives                  299160         151370
11-1021   General and operations managers   1655410        103780
11-1031   Legislators                       61110          33880
11-2011   Advertising and promotions managers 36300        91100
11-2021   Marketing managers                165240         113400
11-2022   Sales managers                    322170         106790
11-2031   Public relations managers         47210          97170
11-3011   Administrative services managers  239360         76370
Time taken: 3.163 seconds, Fetched: 10 row(s)
hive>
```

Like SQL, Hive has a number of ways to see metadata about objects. There are a few ways to see the table details. Try executing one or all of the following commands:

```
DESCRIBE sample_07;
DESCRIBE EXTENDED sample_07;
DESCRIBE FORMATTED sample_07;
```

To leave the HiveCL prompt, you simply type exit with a semicolon. This will exit you back to the shell command line. This quick exercise hopefully helped to show that you will have plenty of options in Hive to view and manipulate data, which should be familiar to anyone familiar with SQL. Facebook created Hive to abstract Java MapReduce from business analysts and make Hadoop accessible to those familiar with SQL and who are most responsible for viewing the data and extracting valuable analytical insights. Since the time Hive was created, it has grown significantly in its performance capabilities as well as its breadth of SQL syntax. Hundreds of companies use Hive today as their primary schema on Hadoop for all their analytics.

This chapter was designed to provide you with a high-level overview of what you can expect to see and accomplish in Hive. The purpose is to give you the entry point into the Hive environment. In later chapters, we will dig deeper and explore the full functionality of Hive. On the surface Hive looks simple. It allows you to quickly begin executing standard SQL syntax against raw structured and semi-structured data, but Hive is much more than that. Hive is highly adaptable and can read numerous file types as well as generate new storage files for near real-time query performance. Many of your existing analytic tools can access the Hive tables you create as if they were accessing a traditional relational database. Users have no idea that the tables they are querying are actually CSV, JSON, XML, or any number of different file types.

Hive is the de facto standard and most widely used SQL-on-Hadoop tool. Hive exists completely in open source and is being continually developed and improved on by Committers across a diverse range of companies. As you begin your journey with Hive, you will find it to be an amazing tool for its simplicity as well as for the complex analytic operations it can achieve.

# CHAPTER 3

■ ■ ■

# Hive Architecture

This chapter digs deeper into the core Hive components and architecture and will set the stage for even deeper discussions in later chapters. Here you will see what makes Hive tick, and what value its architecture provides over traditional relational systems. Make no mistake about it, Hive is complicated but its complexity is surmountable and will be familiar to those who make a living accessing data. Keep in mind too that, like any software development project, Hive is constantly changing and changing fast. Competition in the SQL-on-Hadoop space is driving community innovation at a phenomenal rate. This chapter helps you navigate the core of Hive and aids you in the ride.

## Hive Components

Hive is not a standalone tool and relies on various components for storing and querying data. Within the Hadoop ecosystem, Hive is considered a client data access tool. Data access requires a compute, storage, management, as well as a security framework. Figure 3-1 shows a high-level diagram of these various components.

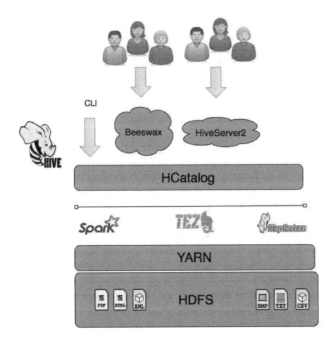

***Figure 3-1.*** *Hive components*

© Scott Shaw, Andreas François Vermeulen, Ankur Gupta, David Kjerrumgaard 2016
S. Shaw et al., *Practical Hive*, DOI 10.1007/978-1-4842-0271-5_3

As mentioned previously, Facebook developed Hive to abstract away the complexities of writing MapReduce, aka Java. This approach overcame serious Hadoop adoption and access barriers, but although Hive smelled like SQL and tasted like SQL, it still was not SQL, especially in regards to processing speed. Under-the-hood Hive queries still ran as MapReduce jobs. MapReduce is batch processing while SQL is an interactive processing language.

---

■ **Note** What is considered batch, interactive, and real-time is somewhat up for debate. The standard definition for interactive is anything that can run in around two seconds. Batch tends to run much longer while real-time is much quicker. The best explanation I've heard is to consider interactive "human time" and real-time "machine time"—think sensor data streaming. Ultimately each person and company will need to determine SLAs.

---

Early on, this mix of what should be an interactive SQL query experience turned into a "wait a day until it completes" experience, which frustrated traditional business intelligence professionals. The fact that the query would potentially run against petabytes of data was little consolation to end users who just wanted interactive data analytics. You can horizontally scale your cluster for additional compute resources and speed up processing, but that would not be a long-term approach and strategy for increasing Hadoop adoption.

Other Hadoop distributors saw the greater need for interactive querying and began developing their own implementations. These include Cloudera's Impala, Pivotal's HAWQ (now Apache HAWQ and also Hortonworks HDB powered by Apache HAWQ), MapR's Drill, Google's BigQuery, IBM's Big SQL, Actian's Vortex, and Jethro SQL, to name a few. Even now the field continues to grow as other processing engines like SparkSQL design their own unique version of SQL on Hadoop. The original Hive on MapReduce open sourced by Facebook required a much needed reengineering to provide similar, competitive functionality to the proprietary offers. This prompted the Stinger and the Stinger.next initiative. As of Hive 1.2.1 you have a choice to run Hive as batch using MapReduce, or as interactive using Tez, or leverage in-memory processing using Spark. Tez is the default execution engine. We will look a bit closer at Tez later in this chapter as well as in subsequent chapters.

# HCatalog

A key component you will need to be familiar with and one we will discuss frequently in this book is HCatalog. When we refer to the concept of schema-on-read versus schema-on-write, HCatalog is what facilitates the schema-on-read. Although usually talked about as a component separate from Hive, HCatalog and Hive are inseparable. When you create a Hive table you create a structure in HCatalog. HCatalog facilitates sharing schemas across various Hadoop components. HCatalog provides a number of key benefits:

- Provides a common schema environment for multiple tools

- Allows for connectors to tools to read data from and write data to Hive's warehouse

- Lets users share data across tools

- Creates a relational structure to Hadoop data

- Abstracts away the how and where of data storage

- Hides schema and storage changes from users

By having HCatalog as the schema metalayer for your tools means when you create a Hive table or use Pig, you do not have to be concerned about where the data is stored or the data format it is stored in. Additionally, you only need to create the table definition once and you can access it using both Pig and Hive.

When you issue a `CREATE TABLE` statement (this should look familiar to anyone working with relational databases) in Hive, such as:

```
CREATE TABLE customers (
        customerid      int,
        firstname       string,
        lastname        string
)
STORED AS orcfile;
```

This statement creates a table definition in the Hive Metastore. For now, do not be concerned with the `STORED AS` clause. This will be discussed in a later chapter. The definition could also contain partitioning information to help with performance, free text comments describing the table, or a directive on whether or not the table is external or internal to Hive. The raw data in HDFS forming the content of the table remains unchanged, but HCatalog applies a structured metalayer defining the data format and data storage. The HCatalog definition resides outside of HDFS. Figure 3-2 shows the database options for the Hive Metastore.

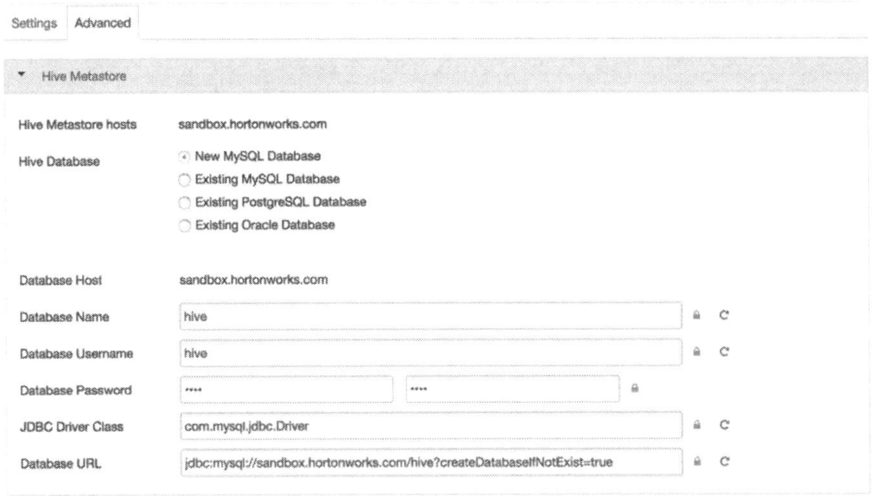

***Figure 3-2.*** *Hive Metastore options*

The Hive database options are MySQL (default), PostgreSQL, and Oracle. Many organizations will choose a database such as Oracle for the Hive repository because the Oracle environment may already provide for security, backup and recovery, and high availability. Local metastore repository will be fine for development environments. For production environments, you will want your Hive Metastore to be secure and protected from failure since it will contain all your table definitions. Keep in mind the files for Hive are stored on HDFS, but the metadata defining the schema for these files exists in a relational database outside of HDFS—either on another server or somewhere on the local Linux filesystem. If you choose Oracle, you will need your Oracle DBA to provide the Oracle JDBC driver as well as access to whatever account you choose in the Hive Metastore settings. Refer here to documentation about installing Hive to a non-default database: `http://docs.hortonworks.com/HDPDocuments/Ambari-2.2.1.0/bk_ambari_reference_guide/ content/_using_non-default_databases_-_hive.html`.

■ **Note** Do not be too concerned about the size of the HCatalog database. Some of the largest Hive implementations only use a couple of terabytes. These are extreme cases where the size of data managed under Hive is in the 100s of petabytes. In most situations, allocating a few gigabytes of space should be plenty.

HCatalog is essentially an abstraction layer between data access tools such as Hive or Pig and the underlying files. In addition, HCatalog provides for an easy separation between those more familiar with the operational aspects of the infrastructure and those more familiar with the LOB (line of business) and the corporate data. Table 3-1 illustrates this process.

*Table 3-1.* *Roles That HCatalog Helps to Facilitate*

| User | Job Function | Activity and Tools |
|---|---|---|
| | User A is responsible for general cluster administrations. He will move data into HDFS, maintain security, and make sure data is available. | Any number of streaming or file copy features. These features may have manual and automated file ingress capabilities. |
| | User B is responsible for cleansing the data and\or creating Hive tables in HCatalog. She is knowledgeable of file formats and general Hive optimization techniques. | She will use Hive to create HCatalog tables. She may also use Pig as an ETL tool to cleanse and modify the data and move it into HCatalog. |
| | User C sees the tables in Hive or in another third-party analytic application and uses them to analyze the data to gain business insight. | Any number of third-party tools can be used to access HCatalog tables. HCatalog accepts ODBC as well as JDBC connections. |

The end user has little concern with how the data is stored, where it is stored, or even how it is specifically schematized. All User C cares about is whether the data is available to the analytic tools at their disposal and whether the data is correct. User A is your traditional operations professional and User B is a traditional ETL\SQL developer. Our guess is, as a reader of this book, you fall into the User B or User C camp, although there is still value if you happen to be in charge of operating the cluster and moving data into the system. Much of the dirty work behind Hive is wrangling data from multiple files coming from various data sources and interpreting these into a loose structure or schema. Later chapters will discuss many of these options and you will find that much of the data wrangling has been done for you.

# Hiveserver2

As beneficial as Hive was at providing a SQL abstraction layer for running MapReduce, there were still some major limitations. One limitation was the ability for clients to connect to the metastore using standard ODBC and JDBC connections. This is something we take for granted in traditional relational database systems. The open source community addressed this limitation by creating the Hive server. Hive server allowed clients to access the metastore using ODBC connections. With Hive server, clients can connect to HCatalog with business intelligence applications like Excel or productivity applications like Toad or SQuirreL.

There were still limitations with Hive server. Primarily, the limitations included user concurrency restrictions as well as security integration with LDAP. Each of these components were solved with the implementation of Hiveserver2. The HiveServer2 architecture is based on a Thrift Service and any number of sessions comprised of a driver, compiler, and executor. The metastore is also a key component of HiveServer2. Figure 3-3 shows a high-level diagram of the HiveServer2 basic architecture.

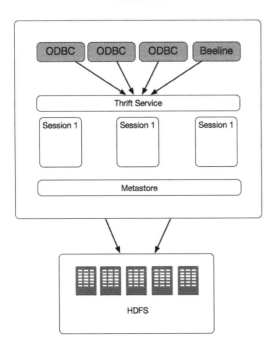

*Figure 3-3.* *HiveServer2 architecture*

Hiveserver2 supports Kerberos, custom authentication, as well as pass-through LDAP authentication. All connection components—JDBC, ODBC, and Beeline—have the ability to use any one of these authentication methods. In addition, HiveServer2 can function in either HTTP mode or TCP (Binary) mode. HTTP mode is useful if you need HiveServer2 to act as a proxy or utilize load balancing. You can access the configurations settings for HiveServer2 in Ambari under the Hive service and Advanced configuration options. Figure 3-4 shows some of these settings. To switch from TCP mode to HTTP mode, you alter the `hive.server2.transport.mode` setting from `binary` to `http`.

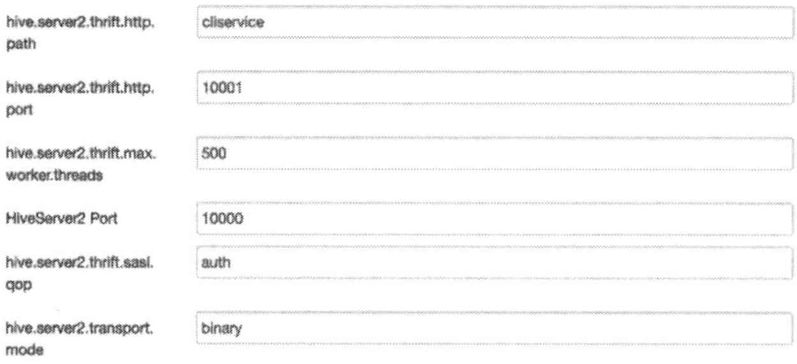

| hive.server2.thrift.http.path | cliservice |
|---|---|
| hive.server2.thrift.http.port | 10001 |
| hive.server2.thrift.max.worker.threads | 500 |
| HiveServer2 Port | 10000 |
| hive.server2.thrift.sasl.qop | auth |
| hive.server2.transport.mode | binary |

***Figure 3-4.*** *HiveServer2 settings*

When connecting to Hive via ODBC, you need to download the appropriate ODBC driver. Many companies and distributions provide their own ODBC connection drivers. Some may be more performant than others. For general purposes, for example when connecting Microsoft's PowerBI to Hive tables, downloading and configuring a Hadoop distributor's ODBC is sufficient. Hortonworks provides various drivers in their add-on section at `http://hortonworks.com/downloads/#data-platform`. Cloudera also offers both ODBC and JDBC drivers on their download site at `http://www.cloudera.com/downloads.html`. Once they are downloaded, you can configure the driver through normal ODBC connection wizards. Figure 3-5 shows an example configuration for using the ODBC driver for Windows.

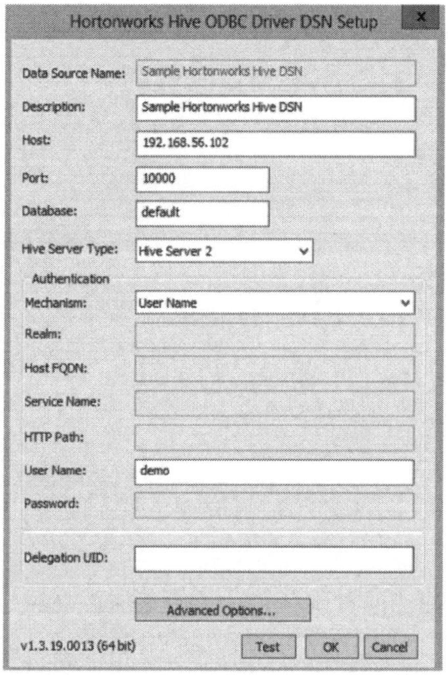

***Figure 3-5.*** *Sample ODBC connection for Windows*

Hiveserver2, introduced in Hive .11 through HIVE-2935 (`https://issues.apache.org/jira/browse/HIVE-2935`), represented a big step in facilitating application access into Hive. It provided for greater concurrency, security, and remote access. As you explore and continue to use the full features of Hive, HiveServer2 will be an integral part of your access to data.

# Client Tools

Throughout this book we will access Hive primarily in one of two ways. The first is through the command-line interface (CLI). This is probably the quickest and most flexible way to access Hive. It allows for cutting and pasting code easily, executing HQL files, as well as a less error prone experience, which sometimes manifests itself in the more graphical tools. As mentioned previously, HiveServer2 allows for both ODBC and JDBC connections so almost any SQL tool has the capability of connecting to Hive. If you are more familiar with a tool such as Toad or SQuirreL, feel free to use those.

We will focus on using the Hortonworks sandbox. As of this writing, the latest available sandbox is HDP 2.4. After downloading and starting the VM and setting the root password, you can simply log into the environment using any SSH compatible shell. Windows users sometimes use Putty to connect. This is especially helpful if you have a large amount of nodes in your cluster and need to list them on within a Putty connection. For our purposes, we will only be connecting to the sandbox that runs on a single node. Connecting via SSH is easily done by starting a CLI window and typing the following code

```
ssh root@hortonworks.sandbox.com -p 2222
```

Once connected, you can enter the Hive CLI by typing hive on the command line. You should now notice a hive> prompt. Navigating within the CLI is straight-forward, especially if you are used to other database systems. Keep in mind that Hive was developed based on MySQL so syntax and data types between the two are quite similar. At the prompt, type:

```
show databases;
```

Now type:

```
show tables;
```

Be sure to end all commands with a semicolon. To see a table's column definition, type:

```
describe <table name>
```

For example, to see the columns for the sample_07 table, type:

```
describe sample_07;
```

Executing hiveql commands is similar to executing any SQL command. To run a simple SELECT statement, type:

```
SELECT * FROM sample_07 LIMIT 10;
```

You will be introduced to more functional commands in subsequent chapters.

Another useful way to issue commands via the command line is through a browser shell. By opening any browser and typing sandbox.hortonworks.com:4200, you can access the command line through a browser. Some developers find the browser simpler than opening up another command window. Copying and pasting can be done but it will always prompt you to select a paste from the browser dialog box. We find it useful when demonstrating the Hive CL due to the browser zoom capabilities. Either way, as you work with the Hive CL, you will begin to develop your own personal preferences.

As demonstrated in an earlier chapter, another primary means to accessing Hive is through Ambari views. Ambari itself is a pluggable framework allowing for developers to create views, which can be installed and executed through the Ambari interface. Views are powerful tools for collaboration and for adding functionality to the Ambari environment. Third-party vendors can create Ambari views for managing their unique application as well as businesses creating their own custom views for internal consumption. View development is outside the scope of this book, but if you want to know more, you can get more information at https://cwiki.apache.org/confluence/display/AMBARI/Views.

Hive has its own Ambari View provided out of the box in HDP 2.4. To get to the views you click on what is referred to as the tic-tac-toe box and select the Hive View. Figure 3-6 shows where the view is located. You will find it in the upper-right corner of the Ambari web page.

**Figure 3-6.** *Ambari Hive view*

The Hive view consists of three main sections: Tool Header, Database Explorer, and Query Editor. The Tool Header is where you can access saved queries, query history, user defined functions, as well as upload a table. The Database Explorer is where you can specify which database you want to use for query execution and a list of all the tables within each database. Clicking on a database will expand the contents of the database and clicking on a table will expand to show the columns and data types of the table. This functionality is similar to the show database, show tables, and describe tables commands used in the Hive CLI. Another feature is when clicking on the ▦ icon, the view will automatically execute a SELECT * statement from the table with a limit of 10. This is a quick way to see sample content.

The Query Editor provides quite a bit of functionality to explore. Besides being the place where you create and execute your Hive queries, you can also use it to customize configuration settings on a per query basis, perform data visualization and data profiling, view visual explain plans and Tez DAG execution and, finally, review logs and error messages. Other functionality includes creating multiple worksheets, saving queries, and killing job executions. Figure 3-7 shows the Hive view screen.

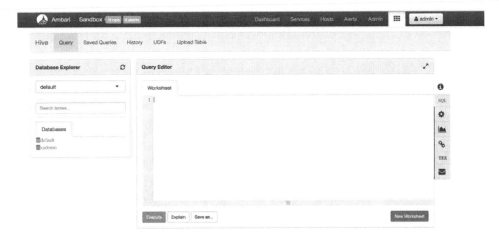

***Figure 3-7.*** *Hive view screen*

The Hive view, as well as all Ambari views, are client-side processes connecting to your Hadoop cluster. As best practice, Ambari runs on an edge node, aka a client node, and connects to your core Hadoop cluster running HDFS. You have the option to set up individual servers running specific views. For example, in one case you may have an Ambari server running the operational dashboard view and another server running the Hive view. This would be useful when you have a large number of operational users as well as a large number of business users. Another option is to have a single Ambari server and give access to views to specific users or groups. To do this, you click on the user button and select Manager Ambari. Figure 3-8 shows the Manage Ambari drop-down option.

***Figure 3-8.*** *Accessing view configuration*

From here, you click on the Views on the left side of the screen and then select Hive View. Figure 3-9 shows how to select the Hive View configuration screen.

***Figure 3-9.*** *Hive view configuration*

It is on the configuration screen where you have the ability to grant permissions to the view to users or groups. These users and groups can be local, LDAP, or Active Directory, depending on your configuration. When you set the permissions, a user will log into Ambari and only see the views to which they have access. All other views will be hidden from them. Figure 3-10 shows the Hive view configuration screen.

***Figure 3-10.*** *Hive view configuration screen*

Hive provides a number of ways to access your data. HiveServer2 provides remote access with security and ODBC and JDBC connections, the CLI provides agile development and control, and the Hive Ambari view provides an easy GUI interface with additional operational functionality. Together, they provide you with the flexibility and the scalability to view and analyze your data.

# Execution Engine: Tez

When Hadoop was conceived, there was only one execution engine from processing data. That engine was MapReduce (MR) and it was a batch operation. It meant that MR had a unique ability to crunch massive amounts of data but processing that data was a monumental task, which not only took up most of your cluster resources, but was also not expected to finish quickly. MR is Java so to access data on Hadoop you had to know Java, specifically how to write Java MR code. As we know, Facebook solved this problem by creating for Hive a SQL abstraction layer for writing MR Java code. This was a huge step in providing access to Hadoop but did nothing for the inherent problems with MR being a batch operation. Users wrote similar SQL code on Hadoop but did not experience anything near the performance they were used to with traditional relational system.

Some early Hadoop distributors solved this problem by creating data access architectures, which accessed data within Hadoop, but processed the data outside of Hadoop. Some of these early SQL-in-Hadoop solutions were based on popular MPP architectures, which utilized parallel processing to gather and execute the data. The goal of many of these operations is to execute in-memory processing for fastest results. Any SQL execution engine tries to execute as much as possible in-memory and avoid costly disk IO operations. The two earlier adopters of this approach include Apache Impala and Apache Hawq. Each follows the same basic pattern of executing SQL commands in parallel across the cluster for maximum distributed processing.

These early SQL-in-Hadoop solutions solved many limitations present with Hive on MapReduce, most importantly performance and ANSI SQL compliance. The problem with early solutions was that they failed to execute on larger data sets. In-memory solutions are performant until the data sets become larger than what can fit into memory. This is because once memory capacity is full, the data needs to spill to disk and you begin hitting IO bottlenecks. The other problem was that they were not Hive or open source. The early SQL solutions were proprietary and included additional costs. Most had only limited connection capabilities to existing Hive metastores. Early Hadoop adopters had been using Hive for years and, instead of looking for a new SQL environment, they would prefer to make Hive better.

The open source community decided to fill the gap by what was marketed as the *Stinger Initiative*. The initiative aimed to provide interactive SQL-in-Hadoop natively in Hive. In order to accomplish this a new engine was required. This new engine was named Tez.

---

■ **Note**    Tez is the Urdu word for swift. Keep in mind, as mentioned, that the open source community is owned and operated by software engineers—not by marketing people—so expect creative naming.

---

Tez became the new paradigm for Hive execution. MapReduce is still supported for Hive execution but Tez is now the default engine when running Hive jobs in Hadoop.

Tez provides a number of advantages over traditional MapReduce. First and foremost, Tez avoids disk IO by avoiding expensive shuffle and shorts while leveraging more efficient map side joins. Tez also utilizes a cost-based optimizer, which helps produce faster execution plans. Combine this with the ORC file format geared toward SQL performance and you have a query engine performing up to 100x faster than native MapReduce. Figure 3-11 shows how Tez is the default engine and the cost-based optimizer is enabled by default.

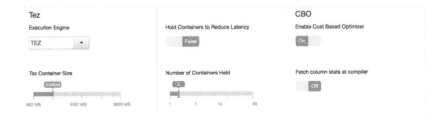

*Figure 3-11.* *Tez and the cost-based optimizer*

Tez and performance tuning will be discussed in detail in Chapter 10, but for now understand that few SQL jobs still utilize MapReduce as the execution engine. There are a number of performance steps you can take to wring the most out of each query. These include but are not limited to using ORC file format and partitioning. Also keep in mind that the current Hive implementation is not an in-memory only process like other data access tools. This is by design since an exclusive in-memory architecture would limit the size of the data sets to only what will fit in memory. Hive is the workhorse of SQL-in-Hadoop and is proven to scale into the petabytes of data.

This chapter focused on some key architecture points in Hive. Throughout the book you will become familiar with these various components and learn how each provides unique value. Hive development is ongoing and fast-paced. We focused on Tez, but Hive can also run on Spark. Again, there are limitations as well as market positioning, all which make the ultimate choice a difficult one. Our advice is to perform your own due diligence. Our focus is on Tez since Tez is an execution engine specifically built from the ground up to work with Hive and provide for interactive SQL latency. The open source community is continuing to work on ever faster and faster data access. The open source Hive architecture provides a flexible foundation for continual develop and innovation to drive SQL analytics that scale well into the future.

# CHAPTER 4

■ ■ ■

# Hive Tables DDL

By now, you know that Hive was created as a means to query the unstructured world of Hadoop without writing complex MapReduce programs. It gives users the ability to write simple queries using the expressiveness of SQL, the language that so many are already familiar with. Hive query language (HiveQL or HQL) is based on ANSI standard SQL, and hence is very easy to understand for anyone familiar with SQL. A user can log in to the Hive command-line interface and start querying the data on HDFS.

Hive provides standard SQL functionality, including many SQL analytical features found in ANSI 2003 and 2011. With each release the Apache community is adding more and more features to HiveQL, bringing it closer to the ANSI SQL. Supporting standard SQL Syntax also extends Hive's usability such that it can easily integrate with existing BI tools such as QlikView, Microstrategy, Microsoft Excel, Power BI, and the like. This integration is done using Hive's ODBC/JDBC driver.

In this chapter we focus on the DDL commands available in HiveQL. We first introduce the concept of a Hive database and data model. We then highlight the different data types that it supports. Most of these data types are quite similar to the world of relational databases, but we also discuss the ones that Hive has inherited directly from programming languages like Java.

Hive has a few different types of tables that allow you to access the structured, semi-structured, and unstructured data effectively. We discuss the concepts including but not limited to creating, altering, dropping tables, columns, partitions, and buckets in this chapter.

## Schema-on-Read

The versatility and power of Hadoop lies in its ability to store and process any kind of unstructured, semi-structured, or structured data. Hive allows the users to create a metadata layer on top of this data and access it using a SQL interface. As much as it is familiar to the end user for its interface, it is different in terms of how it handles the underlying data. Hive does not take control of how data is persisted to disk or its lifecycle. Users can first store any kind of data in HDFS, in its inherent format, and then define metadata to read it independently of the data. Hive makes it easier to manage and process data with a variety of tools with this flexibility. However, since the underlying data can be any format, Hive lets you provide some additional information in the metadata, to explain exactly the nature in which the data stored is formatted. You will notice that in most of the Hive CREATE statements, you provide additional information such as how the underlying data is structured, how the records are defined, and how the fields are separated. Similarly, when you drop external tables in Hive, it will only remove the table's metadata, and not the original data or the HDFS file that contained the data. In most cases you are able to manage the underlying data files directly. The point I am making here is that the user should remember that Hive is not a database; it is a human friendly, familiar interface to query the underlying data files that are stored on HDFS.

© Scott Shaw, Andreas François Vermeulen, Ankur Gupta, David Kjerrumgaard 2016
S. Shaw et al., *Practical Hive*, DOI 10.1007/978-1-4842-0271-5_4

# Hive Data Model

Data models provide a way to organize the data elements and relate them to one another. Hive's data model is quite similar to various relational databases. It consists of a schema of tables, columns, rows, and partitions. These objects are logical units that are defined in the metadata layer called the Hive Metastore. In addition to the common data segments, Hive introduces an additional structure called buckets. The actual data files and directories do not have any information about the data model. The logical units consist of various data types that relate the actual data in the files to columns in the schema. The Hive schema makes the Hadoop data look as if it has familiar rows and columns, whether the underlying data is stored that way or not. This makes the data accessible from common applications that understand SQL languages via ODBC/JDBC.

The metadata repository of Hive also known as Hive Metastore consists of namespaces, object definitions, and the details of underlying data. As of today a Hive Metastore is created in an RDBMS, as it is quite critical to have fast access to this information.

## Schemas/Databases

The concept of databases in Hive is slightly different from what you are probably already familiar with from the RDBMS world. A Hive schema or database is essentially a namespace that holds metadata information for a set of tables. A schema and a database are synonyms in terms of Hive. At the filesystem level, it is a directory under which all internal tables that belong to that namespace are stored. Hive also has a concept of external tables in which the files might exist in other locations in HDFS.

All the data managed by Hive gets stored under a top-level directory defined using the `hive.metastore.warehouse.dir` parameter in the `hive-site.xml` file. The default value of this parameter in the Hortonworks sandbox installation is `/apps/hive/warehouse`. The administrator can change this parameter to another location on HDFS. When you install Hive for the first time, it creates a default database called `default`, which itself does not have its own directory. All the internal tables that you create in the default database are stored under the top-level directory called `hive.metastore.warehouse.dir` in their respective subdirectories. However, all external tables the data exist in other directories in HDFS and the relative locations for these directories are stored in the Hive Metastore.

## Why Use Multiple Schemas/Databases

Prior to the addition of the concept of databases in Hive, all user objects were created in a single namespace. Creating multiple schemas allows users to create objects in different namespaces. Hence it allows for logical grouping of various objects. You can also assign different properties to different database. As an example, you can set different owners for different databases. You can also set different warehouse directories for different databases. From the security perspective, you can grant permissions on all objects in a namespace to a role/user.

## Creating Databases

You can create a database in Hive using the `CREATE DATABASE` command. A simple example of this command is

```
CREATE DATABASE shopping;
```

This command will create a new namespace called `Shopping` in the Hive Metastore. In this example since we have not specified a location for this database on HDFS, it will create a directory called `SHOPPING.db` under the default top-level directory defined in `hive.metastore.warehouse.dir`.

The complete syntax of the CREATE DATABASE command is

```
CREATE (DATABASE | SCHEMA) [ IF NOT EXISTS ] database_name
[ COMMENT database_comment ]
[ LOCATION hdfs_path ]
[ WITH DBPROPERTIES (property_name = property_value,...) ] ;
```

Here is an example using the complete syntax

```
CREATE DATABASE  IF NOT EXISTS  shopping
COMMENT 'stores all shopping basket data'
LOCATION '/user/retail/hive/SHOPPING.db'
WITH DBPROPERTIES ('purpose' = 'testing') ;
```

This command will create a new namespace called shopping and a directory called /user/retail/hive/shopping.db. Using the WITH DBPROPERTIES clause, you can assign any custom properties to a database. You can view these properties using the DESCRIBE DATABASE EXTENDED command as follows:

```
hive> DESCRIBE DATABASE EXTENDED shopping;
OK
shopping stores all shopping basket data
        hdfs://sandbox.hortonworks.com:8020/user/retail/hive/SHOPPING.db root USER
        {purpose=testing}
Time taken: 0.295 seconds, Fetched: 1 row(s)
```

---

■ **Note**  The key point to note in the CREATE database syntax is that the command allows you to specify a location to store the data for the database in a particular location. Hive allows database directories to be created in other locations that are not under the top-level directory specified for the database.

---

## Altering Databases

Once you have created a database, you can modify its metadata properties (DBPROPERTIES) or OWNER using the ALTER DATABASE command as follows:

```
ALTER DATABASE shopping
SET DBPROPERTIES ('department' = 'SALES');
```

## Dropping Databases

You can drop a Hive database using the DROP DATABASE command.

```
DROP DATABASE database_name [RESTRICT|CASCADE];
```

For example:

```
DROP DATABASE shopping CASCADE;
```

The use of CASCADE in this command is optional and allows you to drop a database with existing tables. This command will drop all internal and external tables that belong to the shopping database.

The default behavior of the DROP DATABASE command is RESTRICT, which means if there are any tables in the database, the command will fail.

## List Databases

You can view the list of all databases in the Metastore using

```
SHOW DATABASES [ LIKE 'identifier_with_wildcards' ];
```

For example, SHOW DATABASES LIKE 'S*' will list the shopping database.

# Data Types in Hive

The data types in Hive can be categorized as primitive and complex data types. These data types are implemented in Java. Before we go into the details of complex data types, lets look at the supported primitive data types.

## Primitive Data Types

Just like relational databases, each column value in Hive has a data type, which has constraints and a valid range of values. The behavior of these data types is similar to the underlying data types in Java in which they are implemented. The various types of primitive data types in Hive are as follows:

- Numeric types—Store positive and negative exact and floating-point numbers

- Date/time types—Store temporal values

- Character data types—Store alphanumeric data in strings

- Boolean—True or false

- Binary—Variable length array of binary data

A complete list of primitive data types is very well documented on the Apache web site. You can visit the following link if you require details of any type of primitive data types in Hive: https://cwiki.apache.org/confluence/display/Hive/LanguageManual+Types.

## Choosing Data Types

Hive has a large variety of primitive data types, hence it is crucial that you use the right data type while creating tables. The data types vary in the sense that some of them more restrictive as they have a fixed length, for example, VARCHAR. Historically, while dealing with relational databases, it is more common to use data types with defined length to ensure data integrity. In case of Hadoop, you will often be dealing with various types of data and sometimes you won't know much about the data that will be pushed into the system, hence this restrictive data type approach may not always work. If the data type is too restrictive, Hive will truncate the data to the limit of the defined column width without any warnings. Therefore, it is recommended that you don't choose very restrictive data types while creating tables in Hive.

For example, creating a table with STRING column provides much more flexibility than creating it with VARCHAR(25).

# Complex Data Types

Apart from the primitive data types discussed, Hive also contains few data types that are usually not found in relational databases. These consist of more than one element of primitive data types and are internally implemented using native serializers and deserializers. They allow you to store the data in collection format without having to break it into further individual fields, as you would do in a normalized schema in a relational database. But since Hadoop allows you store any kind of data into its filesystem and read it using schema-on-read, the traditional rules of normalization don't always apply to the underlying data. The complex data types, or collections as they are generally called, are quite useful to map real-world data to a schema layer.

Hive has the following four complex data types:

- Arrays

- Maps

- Structs

- Unions

## Arrays

An array in Hive is an ordered collection of data elements of a similar data type. These elements are represented by sequential subscript values starting from 0. You can access these elements using their corresponding subscript in square brackets. Unlike the arrays in programming languages like Java, you do not define a maximum number of elements in a Hive array.

For example, you can declare an ITEMS array to hold string values as follows:

```
ITEMS ARRAY<"Bread" , "Butter" , "Organic Eggs">
```

Since this collection of strings has a defined order or sequence, these strings can be accessed via a zero-based index.

```
ITEMS[0] returns "Bread"
ITEMS[2] return "Organic Eggs"
```

## Maps

A map is an unordered collection of key-value pairs in Hive. The keys in a map are one of the primitive data types discussed previously. The values, however, can be of any data types that Hive supports, including complex data types. Unlike the arrays where you can access the elements using subscript, the elements of a Map data type are accessed using the keys.

For example, you can declare a Basket collection containing items and their quantities as follows:

```
Basket MAP<'string','int'>
Basket MAP<"Eggs",'12'>
```

You can print a value of quantity by specifying its corresponding Item in the Map function.

```
Basket("Eggs") returns 12.
```

## Structs

Hive structs are similar to structures in some programming languages, such as C. A struct is an object that contains various fields that can be of any data type.

For example, you can declare a customer's address record using the following STRUCT definition:

```
address STRUCT<houseno:STRING, street:STRING, city:STRING, zipcode:INT, state:STRING,
country:STRING>

address <"17","MAIN ST", "SEATTLE", 98104, "WA","USA">
```

You can access the field of a STRUCT using dot notation. In the previous example, the ZIP codes of various addresses can be accessed using `address.zipcode`.

## Unions

A union provides a way to store elements of different data types in different rows of the same field. This is quite useful when the underlying data of a field is not homogenous.

For example, if the customer's contact details are present in the data file but they consist of a single or multiple phone numbers or single or multiple e-mail addresses, you can declare a `contact` variable to store such information as follows.

```
contact UNIONTYPE <int, array<int>, string, array<string>>
```

# Tables

Now that you are already familiar with various data types in the world of Hive, let's look at how these can be used to read data. A Hive data model contains a logical row/column view of data referred to as a *table*. Just like with relational databases, a Hive table consists of a definition on a two dimensional view of the data. However, the data exists independently of the table. The data in a Hive table exists in an HDFS directory and the definition of the table is stored in a relational database store called HCatalog. There are some key differences between the tables in Hive and relational databases:

- The data in a Hive table is loosely coupled with its definition. In relational databases, dropping a table removes its definition and the underlying data from the storage. However, in Hive, if you define a table as an external table, the table definition will be dropped independently of dropping the underlying data.

- A single data set in Hive can have multiple table definitions.

- The underlying data in a Hive table can be stored in a variety of formats. We will discuss some of these file formats in Chapter 7, "Querying Semi-Structured Data".

The separation of actual data from the schema is one of the key value propositions of Hadoop over relational systems. Hadoop allows you to load data even before any schema exists. Once the schema is created, you can modify the schema and determine how it maps to the underlying data in a matter of seconds. Performing such an operation in a relational database requires changes to every row of the table and is not as simple. The Hive schema is just a metadata map, which makes it easy for humans and apps that understand standard SQL to view the underlying data.

## Creating Tables

You can create tables in Hive using the CREATE TABLE statement. Hive's version of CREATE TABLE is quite similar to standard SQL. However it provides various options to add to the versatility of managing various types found in the world of big data. Remember that not all data that we access and manage using Hive is stored natively as rows and columns. It's the configuration specified during the creation of a table that defines how hive will interpret the underlying data, stored as HDFS data files. Hive has many built-in data format interpreters, or *SerDes* as they are called in Hive's language. Hive also allows you to define your own serializer-deserializers and just plug them into a CREATE TABLE statement to enable Hive to understand the format of your data. SerDes are discussed in more detail in Chapter 7. For now, let's look at a simple CREATE TABLE statement.

```
CREATE EXTERNAL TABLE customers (
fname           STRING,
lname           STRING,
address         STRUCT <HOUSENO:STRING, STREET:STRING, CITY:STRING, ZIPCODE:INT,
                STATE:STRING, COUNTRY:STRING>,
active          BOOLEAN,
created         DATE
LOCATION '/user/demo/customers');
```

This CREATE TABLE example uses some of the data types discussed earlier. Unless you changed the active database before running this command, it will create a customers table in the default database. You can also create a table directly in a database by prefixing the table name with "the database name". Here is an example of this:

```
CREATE EXTERNAL TABLE retail.customers (
fname           STRING,
lname           STRING,
address         STRUCT <HOUSENO:STRING, STREET:STRING, CITY:STRING, ZIPCODE:INT,
                STATE:STRING, COUNTRY:STRING>,
active          BOOLEAN,
created         DATE)
COMMENT "customer master record table"
LOCATION '/user/demo/customers/';
```

## Listing Tables

You can list the existing tables using the SHOW TABLES command. Let's see the current list of tables in the RETAIL database.

```
hive> SHOW TABLES IN retail;
OK
customers
Time taken: 0.465 seconds, Fetched: 1 row(s)
```

If you have many tables in a database, you can search for specific tables using wildcards.

55

## Internal/External Tables

Hive tables can be created as internal or external. The type of Hive table determines how the data is loaded, stored, and controlled by Hive.

## External Tables

External tables are created using EXTERNAL keywords in the CREATE TABLE statement. This is the recommended table type for all production deployments of Hadoop. This is because in most cases the underlying data will be used for multiple use cases. Even if its not, it should not be dropped when the table definition is dropped. So in case of external tables, Hive does not drop the data from the filesystem as it does not have control over it. You use external tables in the following cases:

- When you want to drop the table definitions without worrying about deleting the underlying data.

- When the data stored in filesystem others than HDFS. For example, you can use s3 in case of Amazon or WASB in case of Microsoft Azure to store data and access that data from multiple clusters.

- You want to use a custom location to store the table data.

- You are not creating a table based on another table (CREATE TABLE AS SELECT).

- Data will be accessed by multiple processing engines. For example, you want to read the table using Hive but also want to use it in a Spark program.

- You want to create multiple tables' definitions over the same data set. It is important that if you have multiple table definitions, dropping one of them should not delete the underlying data.

## Internal or Managed Tables

An internal table in Hive refers to a table whose data is managed by Hive. This means when you delete an internal table Hive will also delete its underlying data. These tables are not used very often in Hadoop as in most environments, the data in the filesystem needs to remain even after the table is dropped. Since the data and the metadata in Hive are not tied together, this allows for the underlying data to be used with other tools/processing paradigms. You use internal tables in the following cases:

- When the data stored is temporary.

- When the only way the data is accessed is through Hive and you want Hive to completely manage the lifecycle of the table and the data.

---

▓ **Note**    Remember that you can always modify/delete the underlying data directly on HDFS even when the tables are internal/managed. This is because Hive does not have full control over the underlying data. The differentiation between internal and external table data control is based on how the data is deleted through Hive, such as when you drop the table.

---

# External/Internal Table Example

We will now walk through a basic example to demonstrate some of the differences between external and internal tables.

Load a file to HDFS and verify it.

```
hadoop fs -put /tmp/states.txt /user/demo/states/
hadoop fs -ls /user/demo/states
Found 1 items
-rw-r--r--   3 demo hdfs         58 2016-07-02 21:02 /user/demo/states/states.txt
```

Now, let's first create an internal table to access the data in the states.txt file.

```
hive> CREATE TABLE states_internal (state string) LOCATION '/user/demo/states';
OK
Time taken: 8.918 seconds
```

Hive will only output the time taken to process this command. We can see the table definition as follows:

```
hive> DESCRIBE FORMATTED states_internal;
OK
# col_name              data_type                comment

state                   string

# Detailed Table Information
Database:               default
Owner:                  demo
CreateTime:             Sat Jul 02 21:05:14 UTC 2016
LastAccessTime:         UNKNOWN
Protect Mode:           None
Retention:              0
Location:               hdfs://sandbox.hortonworks.com:8020/user/demo/states
Table Type:             MANAGED_TABLE
Table Parameters:
        COLUMN_STATS_ACCURATE    false
        numFiles                 1
        numRows                  -1
        rawDataSize              -1
        totalSize                58
        transient_lastDdlTime    1467493514

# Storage Information
SerDe Library:          org.apache.hadoop.hive.serde2.lazy.LazySimpleSerDe
InputFormat:            org.apache.hadoop.mapred.TextInputFormat
OutputFormat:           org.apache.hadoop.hive.ql.io.HiveIgnoreKeyTextOutputFormat
Compressed:             No
Num Buckets:            -1
Bucket Columns:         []
Sort Columns:           []
Storage Desc Params:
        serialization.format    1
Time taken: 0.559 seconds, Fetched: 31 row(s)
```

You can see from this output that it's a MANAGED_TABLE and also its location.

We can query the data from this table as follows:

```
hive> SELECT * FROM states_internal;
OK
california
ohio
north dakota
new york
colorado
new jersey
Time taken: 1.834 seconds, Fetched: 6 row(s)
```

You can also create an internal table without specifying any location. In such a case, Hive will store its data under the default Hive directory.

Now, let's add another file to the /user/demo/states directory.

```
hadoop fs -put /tmp/morestates.txt /user/demo/states/
```

We will now query the data in states_internal table again.

```
hive> SELECT * FROM states_internal;
OK
new mexico
hawaii
oregon
south dakota
california
ohio
north dakota
new york
colorado
new jersey
Time taken: 7.32 seconds, Fetched: 10 row(s)
```

As you can see from this output, we can now query the data from both files under the /user/demo/states directory. This is because when we created the table we specified the directory as the location.

Now let's create an external table on the same data.

```
hive> CREATE EXTERNAL TABLE states_external (state string) LOCATION '/user/demo/states';
OK
Time taken: 2.57 seconds
```

Let's take a look at its schema.

```
hive> DESCRIBE FORMATTED states_external;
OK
# col_name              data_type               comment

state                   string
```

```
# Detailed Table Information
Database:               default
Owner:                  hdfs
CreateTime:             Sat Jul 02 21:19:31 UTC 2016
LastAccessTime:         UNKNOWN
Protect Mode:           None
Retention:              0
Location:               hdfs://sandbox.hortonworks.com:8020/user/demo/states
Table Type:             EXTERNAL_TABLE
Table Parameters:
        EXTERNAL                TRUE
        transient_lastDdlTime   1467494371


# Storage Information
SerDe Library:          org.apache.hadoop.hive.serde2.lazy.LazySimpleSerDe
InputFormat:            org.apache.hadoop.mapred.TextInputFormat
OutputFormat:           org.apache.hadoop.hive.ql.io.HiveIgnoreKeyTextOutputFormat
Compressed:             No
Num Buckets:            -1
Bucket Columns:         []
Sort Columns:           []
Storage Desc Params:
        serialization.format    1
Time taken: 0.469 seconds, Fetched: 27 row(s)
```

We can query the data in this table as follows:

```
hive> SELECT * FROM states_external;
OK
new mexico
hawaii
oregon
south dakota
california
ohio
north dakota
new york
colorado
new jersey
Time taken: 7.363 seconds, Fetched: 10 row(s)
```

Now, we have two tables on the same data set. This way you can create multiple tables over the same data.

Let's create another external table on the same data set.

```
hive> CREATE EXTERNAL TABLE states_external2 (state string) LOCATION '/user/demo/states';
OK
Time taken: 2.548 seconds
```

We can now query the same data using any of the three tables that we created in this example. Now let's see what happens when we drop the tables. We will drop the second external table.

```
hive> DROP TABLE states_external2;
OK
Time taken: 0.656 seconds
```

Let's see if we can still query the data using the other two tables.

```
hive> SELECT * FROM states_internal;
OK
new mexico
hawaii
oregon
south dakota
california
ohio
north dakota
new york
colorado
new jersey
Time taken: 0.546 seconds, Fetched: 10 row(s)

hive> SELECT * FROM states_external;
OK
new mexico
hawaii
oregon
south dakota
california
ohio
north dakota
new york
colorado
new jersey
Time taken: 0.557 seconds, Fetched: 10 row(s)
```

As you can see, dropping an external table doesn't affect the underlying data. We will now drop the internal table.

```
hive> DROP TABLE states_internal;
OK
Time taken: 0.571 seconds
```

Let's try to query the data using the external table.

```
hive> SELECT * FROM states_external;
OK
Time taken: 0.545 seconds
```

Since Hive controls the INTERNAL table and the underlying data, when we dropped the states_internal table, Hive also deleted the underlying data. This is why when we tried to query the data from states_external, there is no output.

# Table Properties

You can also specify some table-level properties while creating a table or by altering a table using the TBLPROPERTIES clause. Hive has some predefined essential properties for tables, which you can define some table level configuration that Hive uses to manage the table. However, you can also define some custom properties using a key-value format to store some table-level metadata or additional information about the table.

Here are some of the important table-level properties in Hive.

- last_modified_user

- last_modified_time

- immutable

- orc.compress

- skip.header.line.count

The first two properties in this list are managed and populated by Hive automatically. As their names suggest, these are used by Hive to store the last modified user and time information in the metastore.

When the immutable property is set to TRUE, no new rows can be inserted in a table if it already contains some data. If you try to insert data into an immutable table, you get the following error:

```
hive> INSERT INTO test1 VALUES ('bacon');
FAILED: SemanticException [Error 10256]: Inserting into a non-empty immutable table is not
allowed test1
```

The orc.compress property is used to specify the compression algorithm used for ORC-based storage. We will discuss the ORC files further in this chapter in the section entitled "ORC File Format".

The skip.header.line.count property is one of the most important properties for an external table in Hive. In most production environments, this property is used quite frequently. When dealing with the real-life data, you will often find that the header row in data files is a perpetual headache. Using this property, you can skip a header row from the underlying data files.

Let's see how we can use this property using an example.

We will first copy a file to HDFS.

```
hadoop fs -put /tmp/states3.txt /user/demo/states3
```

Let's also list the data from this file.

```
hadoop fs -cat /user/demo/states3/states3.txt
STATE_NAME
----------
california
ohio
north dakota
new york
colorado
new jersey
```

As you can see from this output, the data file contains two header rows. We will now create an EXTERNAL table with the skip.header.line.count property to read the data from this file without the headers.

```
hive> CREATE EXTERNAL TABLE states3 (states string) LOCATION '/user/demo/states3'
TBLPROPERTIES("skip.header.line.count"="2");
OK
Time taken: 9.0 seconds
```

Let's query the data from this table.

```
hive> SELECT * FROM states3;
OK
california
ohio
north dakota
new york
colorado
new jersey
Time taken: 0.553 seconds, Fetched: 6 row(s)
```

Without this property, Hive interprets the first two header rows as regular strings and will show them in the output of the SELECT command.

## Generating a Create Table Command for Existing Tables

You can also generate the CREATE TABLE statement for a given table using SHOW CREATE TABLE as follows:

```
hive> SHOW CREATE TABLE states3;
OK
CREATE EXTERNAL TABLE `states3`(
   `states` string)
ROW FORMAT SERDE
   'org.apache.hadoop.hive.serde2.lazy.LazySimpleSerDe'
STORED AS INPUTFORMAT
   'org.apache.hadoop.mapred.TextInputFormat'
OUTPUTFORMAT
   'org.apache.hadoop.hive.ql.io.HiveIgnoreKeyTextOutputFormat'
LOCATION
   'hdfs://sandbox.hortonworks.com:8020/user/demo/states3'
TBLPROPERTIES (
   'COLUMN_STATS_ACCURATE'='false',
   'numFiles'='1',
   'numRows'='-1',
   'rawDataSize'='-1',
   'skip.header.line.count'='2',
   'totalSize'='80',
   'transient_lastDdlTime'='1467497215')
Time taken: 0.37 seconds, Fetched: 18 row(s)
```

## Partitioning and Bucketing

Hive tables can be broken in further logical chunks for ease of management and improving performance. There are few ways you can further abstract data in Hive. See Figure 4-1.

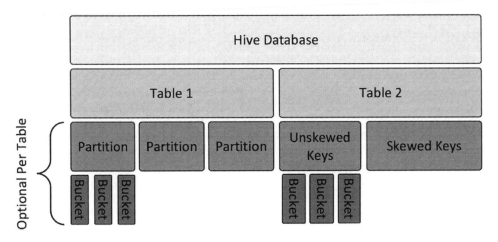

**Figure 4-1.** *Hive data model representation*

## Partitioning

Partitioning is often used in the relational database world to enhance performance and for better management of the data. The concept of partitioning in Hive is no different.

Partitioned tables in Hive have one or more partition keys based on which the data is broken into logical chunks and stored in separate directories. Each partition key adds a level of directory structure to the table storage. Let's look at an example of a customer transaction table with some partitioning keys.

```
CREATE EXTERNAL TABLE retail.transactions (
transdate       DATE,
transid         INT,
custid          INT,
fname           STRING,
lname           STRING,
item            STRING,
qty             INT,
price           FLOAT
)
PARTITIONED BY  (store STRING);
```

The table in this example is partitioned on a string column called STORE that will contain the name of the store. Notice that actual column used in partitioning does not exist in the CREATE TABLE structure. This is different from most relational databases, where you have to specify the partition column or partition key in the actual CREATE TABLE structure as one of the columns of the table. If your data already contains the partition key in the format, it probably doesn't make sense to remove it. You can give it a different name and hide it with a view.

When you query a partitioned table, the value of the partition shows up as the value for the column for all rows in that partition. For example, SELECT * FROM retail.transactions returns the values for the store column, even though that data isn't stored in the data files.

Creating a partitioned table requires that you create the directory structure for the underlying partitions beforehand. In case of internal tables, the partition directories are automatically created when you insert data into a new partition using the INSERT command.

```
INSERT INTO transactions_int PARTITION (store="new york") values ("01/25/2016",101,"A109","M
ATTHEW","SMITH","SHOES",1,112.9);
Query ID = hdfs_20160702224145_28638e82-a6cc-4f9f-9c91-86d4a4fadd39
Total jobs = 1
Launching Job 1 out of 1

Status: Running (Executing on YARN cluster with App id application_1467479265950_0010)

--------------------------------------------------------------------------------
        VERTICES      STATUS  TOTAL  COMPLETED  RUNNING  PENDING  FAILED  KILLED
--------------------------------------------------------------------------------
Map 1 .........   SUCCEEDED      1          1        0        0       0       0
--------------------------------------------------------------------------------
VERTICES: 01/01  [==========================>>] 100%  ELAPSED TIME: 4.28 s
--------------------------------------------------------------------------------
Loading data to table default.transactions_int partition (store=new york)
Partition default.transactions_int{store=new york} stats: [numFiles=1, numRows=1,
totalSize=38, rawDataSize=37]
OK
Time taken: 11.081 seconds

hive> SHOW PARTITIONS transactions_int;
OK
store=new york
```

---

■ **Caution**    If you try to include the partition key column in the actual table definition, you will get "FAILED:
Error in semantic analysis: Column repeated in partitioning columns".

---

## Partitioning Considerations

Hive partitioning can improve the performance of a very specific subset of queries by pruning the partitions that are not required to retrieve the results of the query. This is called partition elimination. Partitioning is also one ways to enable the users to organize the data on HDFS in a more segmented manner that may improve maintainability. If your data is broken into subdirectories, you can either point partitions to the subdirectories or enable recursive partitions to allow a single table access all the subdirectories. If you have subdirectories without one of those options, you will see errors or empty data sets from queries of the Hive tables.

Just as with relational database, if used incorrectly, partitioning can lead to degradation of performance. The key thing with Hive partitioning is not to overpartition. Partitions increase the overhead in both data loading and data retrieval. If you create a very large number of partitions with small chunk of data in each partition, you are more likely to have small files. Large number of small files is generally much slower in Hadoop than fewer, larger files. Some of the best practices to consider when partitioning tables in Hive are as follows:

- Pick a column for partition key with low to medium Number of Distinct Values (NDVs).

- Avoid partitions that are less than 1 GB (bigger is better).

- Tune Hiveserver2 and Hive Metastore memory for large number of partitions.

- When you use multiple columns for partition key, it will create a nested tree of subdirectories for each combination of partition key columns. You should avoid deep nesting as it can cause too many partitions and hence create very small files.

- When insert data using Hive streaming, if multiple sessions write data to same partitions, it can lead to locking.

- You can modify the schema of a partitioned table; however, once the structure is changed, you cannot modify the data in existing partitions

- If you are inserting data to multiple partitions in parallel, you should set `hive.optimize.sort.dynamic.partition` only to `True`.

## Efficiently Partitioning on Date Columns

Date types are usually one of the most common candidates for a partition key. There are many use cases where you might want to partition the data with the date. One common example is if you are loading various log files in HDFS and want to query them using Hive, then you might want to organize the data per day. When creating partitions by date, it is almost always more effective to partition by a single string of YYYY-MM-DD rather than use a multi-depth partition with the year, months, and days all as their own values. The advantage to using the single-string approach is that it allows for more SQL operators to be utilized, such as `LIKE`, `IN`, and `BETWEEN`, which cannot be as easily used if you use nested partitions.

Let's say we have a table A, partitioned by a `DateStamp` string as YYYY-MM-DD. We can run various queries on this table using different SQL operators, as follows:

```
Query to select specific Dates
SELECT * FROM Table A WHERE DateStamp IN ('2015-01-01', '2015-02-03', '2016-01-01');

Querying all dates in a Year
SELECT * FROM TableA WHERE DateStamp LIKE '2015-%';

Querying all dates in a particular month of a year
SELECT * FROM TableA WHERE DateStamp LIKE '2015-02-%';

Querying All Days that start/end with a 5
SELECT * FROM TableA WHERE DateStamp LIKE '%-%-%5';

Querying All Days Between 2015-01-01 and 2015-03-01
SELECT * FROM TableA WHERE DateStamp BETWEEN '2015-01-01' AND '2015-03-01';
```

## Bucketing

Bucketing in Hive is another way to cut data into smaller segments. So far we have seen how partitioning can help organize and access the data efficiently. However, efficient partitioning requires the use of a partition key, which will not lead to a large number of very small partitions. So if you have many different values for the partition key and not many rows for each value of partition key, partitioning may not be the best choice.

Bucketing is more suitable for such cases. Bucketing lets you define the maximum number of buckets for a bucketed column of the table. A partition is a directory in Hive, where the partition key value gets stored in the actual partition directory name and partition key is a virtual column in the table. However, in case of bucketing, each bucket is a file that holds the actual data that is broken on the basis of a hash algorithm. Bucketing does not add a virtual column to the table.

Like partitioning, bucketing has its own advantages, the primary one being performance improvement with various queries. We will look into some of these benefits in next section. If the key used for bucketing is not skewed, you will have a uniform distribution of data. This can be used for performing efficient data sampling.

Here is an example of creating a table with bucketing. Here we are creating a CUSTOMERS table with the CREATED column as a bucketed column; it's broken into 11 buckets.

```
hive> CREATE EXTERNAL TABLE customers (
    > custid          INT,
    > fname           STRING,
    > lname           STRING,
    > city           STRING,
    > state           STRING
    > )
    > CLUSTERED BY (custid) INTO 11 BUCKETS
    > LOCATION '/user/demo/customers';
OK
Time taken: 1.22 seconds
```

Now when you insert data into this table, Hive will use custid in a Hash function to distribute the data into 11 buckets. For some data types, it means that the rows containing the same value of custid will reside in same bucket.

---

■ **Caution**    Set hive.enforcing.bucketing= TRUE. Without this parameter you need to define the same number of mappers as the number of buckets for a table.

---

## Bucketing Considerations

Bucketing is a great feature for efficient sampling and improving performance of some queries; however, it has its own caveats. Skewness is one of the most common problems encountered in real-world data and this can have a major impact in bucketing if it's not handled properly. Choosing the right key for bucketing is also very important.

Following are some of the best practices that you should follow when using bucketing in Hive:

- Choose a bucket key with high number of distinct values. This reduces the chances of skewness.

- Use a prime number for number of buckets.

- If the data in the bucket key is skewed, create separate buckets for skewed values. This can be done using list bucketing.

- Bucketing is most useful for fact tables, which are often joined together.

- The bucket count of the tables that need to be joined must either be the same or a factor of the number of buckets of the other table.

- Choose the number of buckets carefully. Only one CPU core writes to a single bucket so a large cluster can be severely underutilized if the number of buckets is small.

- The number of buckets cannot be changed once the table is created.

- Carefully choose the column for bucketing, as skewness can be introduced by the hash function. String hashing is more prone to this as usually a small subset of characters is used. For example, if the three values in the bucket key contain ABC789, ABC567, and ABC123, but the hashing algorithm only uses first three characters (ABC) for calculating the candidate bucket, all three may end up in same bucket.

- You should aim to get the bucket file sizes of at least 1 GB.

- Enforce bucketing by setting hive.enforce.bucketing=true.

- Map-side joins are faster on bucketed tables than non-bucketed tables. In a map-side join, a mapper processing a bucket of the left table knows that the matching rows in the right table will be in its corresponding bucket, so it only retrieves that bucket, which is a small fraction of all the data stored in the right side table.

- Bucketing also allows you to sort the data in each bucket by one or more columns. This converts map-side joins into sort-merge joins, which are even faster.

## Temporary Tables

As of Hive 0.14, Hive also support temporary tables. Temporary tables hold the data for the life of a session. This is quite convenient for some applications that need to store intermediate data for the life of the processing and then delete it automatically at the end. Unlike internal tables, temporary tables store their data in the user's scratch directory. By default, the scratch directory is /tmp/hive-username. Different users can create a temporary table with the same name in the same namespace as it is created in their private scratch area.

Here is an example that creates a temporary table and views its properties using the DESCRIBE EXTENDED command.

```
hive> CREATE TEMPORARY TABLE states (state STRING);
OK
Time taken: 2.378 seconds
hive> DESCRIBE EXTENDED states;
OK
state                    string

Detailed Table Information       Table(tableName:states, dbName:default, owner:hdfs,
createTime:1467549942, lastAccessTime:0, retention:0, sd:StorageDescriptor(cols:[Field
Schema(name:state, type:string, comment:null)], location:hdfs://sandbox.hortonworks.
com:8020/tmp/hive/hdfs/bf1e3648-d165-47f7-b27e-1e1f488f29f7/_tmp_space.db/d494a62e-
c1a5-4609-a9f6-4a26e656eebb, inputFormat:org.apache.hadoop.mapred.TextInputFormat,
outputFormat:org.apache.hadoop.hive.ql.io.HiveIgnoreKeyTextOutputFormat, compressed:false,
numBuckets:-1, serdeInfo:SerDeInfo(name:null, serializationLib:org.apache.hadoop.hive.
serde2.lazy.LazySimpleSerDe, parameters:{serialization.format=1}), bucketCols:[],
sortCols:[], parameters:{}, skewedInfo:SkewedInfo(skewedColNames:[], skewedColValues:[],
skewedColValueLocationMaps:{}), storedAsSubDirectories:false), partitionKeys:[],
parameters:{}, viewOriginalText:null, viewExpandedText:null, tableType:MANAGED_TABLE, pri
vileges:PrincipalPrivilegeSet(userPrivileges:{hdfs=[PrivilegeGrantInfo(privilege:INSERT,
createTime:-1, grantor:hdfs, grantorType:USER, grantOption:true), PrivilegeGrantInfo(privile
ge:SELECT, createTime:-1, grantor:hdfs, grantorType:USER, grantOption:true), PrivilegeGrant
Info(privilege:UPDATE, createTime:-1, grantor:hdfs, grantorType:USER, grantOption:true),
PrivilegeGrantInfo(privilege:DELETE, createTime:-1, grantor:hdfs, grantorType:USER,
grantOption:true)]}, groupPrivileges:null, rolePrivileges:null), temporary:true)
Time taken: 0.176 seconds, Fetched: 3 row(s)
```

# Altering Tables

You can modify the existing table structures using the ALTER TABLE command. This command is quite similar to the standard SQL ALTER TABLE command and has few different functions in Hive. All options in ALTER TABLE enable you to modify the structures of the tables but they do not modify the data.

Let's look at a few options in ALTER TABLE.

## Renaming Tables

You can rename a table using ALTER TABLE RENAME command. As an example, we will rename our states table to states_old and then view its properties.

```
hive> CREATE EXTERNAL TABLE states (state STRING) LOCATION '/user/demo/states';
OK
Time taken: 1.057 seconds
hive> ALTER TABLE states RENAME TO states_old;
OK
Time taken: 1.211 seconds
hive> DESCRIBE FORMATTED states_old;
OK
# col_name                data_type                comment

state                     string

# Detailed Table Information
Database:                 default
Owner:                    hdfs
CreateTime:               Sun Jul 03 13:03:15 UTC 2016
LastAccessTime:           UNKNOWN
Protect Mode:             None
Retention:                0
Location:                 hdfs://sandbox.hortonworks.com:8020/user/demo/states
Table Type:               EXTERNAL_TABLE
Table Parameters:
        COLUMN_STATS_ACCURATE    false
        EXTERNAL                 TRUE
        last_modified_by         hdfs
        last_modified_time       1467551010
        numFiles                 5
        numRows                  -1
        rawDataSize              -1
        totalSize                213
        transient_lastDdlTime    1467551010
```

```
# Storage Information
SerDe Library:          org.apache.hadoop.hive.serde2.lazy.LazySimpleSerDe
InputFormat:            org.apache.hadoop.mapred.TextInputFormat
OutputFormat:           org.apache.hadoop.hive.ql.io.HiveIgnoreKeyTextOutputFormat
Compressed:             No
Num Buckets:            -1
Bucket Columns:         []
Sort Columns:           []
Storage Desc Params:
        serialization.format    1
Time taken: 0.575 seconds, Fetched: 34 row(s)
```

## Modifying a Table's Storage Properties

You can modify the storage properties of a table using the ALTER TABLE command in Hive. However, the better recommended approach is to extract the CREATE TABLE statement (or pull it out of version control if its stored), drop the table, modify the CREATE TABLE with new storage attributes, and recreate it. In most production environments, table definitions are maintained in version control and doing things this way also maintains a record of the change performed.

## ORC File Format

The ORC file format is designed to reduce the amount of data read from the disk. Many of the new performance optimizations in Hive only work with ORC files, hence for most use cases, it is recommended that the raw data be converted into ORC files. This format is explained in detail in Chapter 9, "Performance Tuning: Hive". In this section, we discuss the steps that we can follow to convert a text file-based external table into an ORC file.

Let's convert our states table into the ORC format and view its properties.

```
hive> CREATE TABLE states_orc STORED AS ORC TBLPROPERTIES("ORC.COMPRESS"="SNAPPY") AS SELECT *
FROM states;
Query ID = hdfs_20160703133105_d38ec632-7250-42ac-bb58-23e2ed2028ec
Total jobs = 1
Launching Job 1 out of 1
Tez session was closed. Reopening...
Session re-established.

Status: Running (Executing on YARN cluster with App id application_1467537169806_0004)

--------------------------------------------------------------------------------
        VERTICES      STATUS   TOTAL  COMPLETED  RUNNING  PENDING  FAILED  KILLED
--------------------------------------------------------------------------------
Map 1 ..........    SUCCEEDED      1          1        0        0       0       0
--------------------------------------------------------------------------------
VERTICES: 01/01  [==========================>>] 100%  ELAPSED TIME: 4.90 s
--------------------------------------------------------------------------------
Moving data to: hdfs://sandbox.hortonworks.com:8020/apps/hive/warehouse/states_orc
Table default.states_orc stats: [numFiles=1, numRows=22, totalSize=364, rawDataSize=2024]
OK
```

```
Time taken: 14.461 seconds
hive> DESCRIBE EXTENDED states_orc;
OK
state                    string
```

Detailed Table Information    Table(tableName:states_orc, dbName:default, owner:hdfs, createTime:1467552677, lastAccessTime:0, retention:0, sd:StorageDescriptor(cols:[FieldSchema(name:state, type:string, comment:null)], location:hdfs://sandbox.hortonworks.com:8020/apps/hive/warehouse/states_orc, **inputFormat:org.apache.hadoop.hive.ql.io.orc.OrcInputFormat, outputFormat:org.apache.hadoop.hive.ql.io.orc.OrcOutputFormat**, compressed:false, numBuckets:-1, serdeInfo:SerDeInfo(name:null, serializationLib:org.apache.hadoop.hive.ql.io.orc.OrcSerde, parameters:{serialization.format=1}), bucketCols:[], sortCols:[], parameters:{}, skewedInfo:SkewedInfo(skewedColNames:[], skewedColValues:[], skewedColValueLocationMaps:{}), storedAsSubDirectories:false), partitionKeys:[], parameters:{numFiles=1, **ORC.COMPRESS=SNAPPY**, transient_lastDdlTime=1467552677, COLUMN_STATS_ACCURATE=true, totalSize=364, numRows=22, rawDataSize=2024}, viewOriginalText:null, viewExpandedText:null, tableType:MANAGED_TABLE)
Time taken: 0.574 seconds, Fetched: 3 row(s)

## Merging a Table's Files

Dealing with small files is a constant challenge in Hadoop, as they consume lot of NameNode entries for metadata. It is always recommended to stitch small files together into bigger ones. If you have a table in the ORC file format with many small files, you can merge them to make optimum use of HDFS metadata space in the NameNode. Do this using the ALTER TABLE command. The HDFS NameNode process maintains the metadata of all files in HDFS.

For Hive tables stored as RCFile or ORCFile, this can be done as follows:

```
ALTER TABLE states CONCATENATE;
```

This command will merge multiple data files into larger files.

The best way to avoid small files is to merge them into files many times the size of the cluster's chunk size, generally many gigabytes or larger, before they land in Hadoop. There are many ways to stitch them together after they land in Hadoop, but since Hadoop doesn't perform well with many smaller files, those stitching processes can themselves be slow.

## Altering Table Partitions

So far we have seen a few options of the ALTER TABLE command to modify a few table properties. You can also use this command to modify table partitions with some additional options.

## Add Partition

You can add new partitions to existing tables using the ALTER TABLE ADD PARTITION command. As new data is loaded into HDFS into subdirectories of an existing external partitioned table, you would need to run this command to plug in the new partitions. This command allows you add one or more partitions based on the same existing partition key to an existing table.

Let's take a look at an example to add a new partition to an existing table. We will first create a directory for the external table and two partitions on HDFS.

```
hadoop fs -mkdir /user/demo/ids
hadoop fs -mkdir /user/demo/ids/2016-05-31
hadoop fs -mkdir /user/demo/ids/2016-05-30
```

Now, we will copy the data to these directories.

```
hadoop fs -put /tmp/2016-05-31.txt /user/demo/ids/2016-05-31/
hadoop fs -put /tmp/2016-05-30.txt /user/demo/ids/2016-05-30/
```

Let's now create the external table and add partitions to it.

```
 hive> CREATE EXTERNAL TABLE ids (a INT) PARTITIONED BY (datestamp STRING) LOCATION '/user/
demo/ids';
OK
Time taken: 1.009 seconds
```

Let's add the partitions to the table now.

```
ALTER TABLE ids ADD PARTITION (datestamp='2016-05-30') location '/user/demo/ids/2016-05-30';
hive> SELECT * FROM ids;
OK
11      2016-05-30
12      2016-05-30
13      2016-05-30
14      2016-05-30
15      2016-05-30
16      2016-05-30
Time taken: 1.011 seconds, Fetched: 6 row(s)
```

Similarly, we can add the other partition to this table.

```
hive> ALTER TABLE ids ADD PARTITION (datestamp='2016-05-31') location '/user/demo/ids/2016-
05-31';
OK
Time taken: 0.438 seconds
hive> SELECT * FROM ids;
OK
11      2016-05-30
12      2016-05-30
13      2016-05-30
14      2016-05-30
15      2016-05-30
16      2016-05-30
1       2016-05-31
2       2016-05-31
3       2016-05-31
4       2016-05-31
5       2016-05-31
6       2016-05-31
Time taken: 0.649 seconds, Fetched: 12 row(s)
```

For an internal table, you can add the new partitions using the MSCK REPAIR TABLE command. Let's look at an example of this. We will first create an internal partitioned table called ids_internal.

```
hive> CREATE TABLE ids_internal (a INT) PARTITIONED BY (datestamp STRING);
OK
Time taken: 2.422 seconds
```

Now let's add a couple of rows in two different partitions.

```
hive> INSERT INTO ids_internal PARTITION (datestamp='2016-05-30') values (1);
Query ID = hdfs_20160703164138_82dfaa1f-e746-4c68-b694-0bb639af2961
Total jobs = 1
Launching Job 1 out of 1

Status: Running (Executing on YARN cluster with App id application_1467537169806_0011)

--------------------------------------------------------------------------------
        VERTICES      STATUS   TOTAL   COMPLETED   RUNNING   PENDING   FAILED   KILLED
--------------------------------------------------------------------------------
Map 1 .........   SUCCEEDED     1         1          0         0         0        0
--------------------------------------------------------------------------------
VERTICES: 01/01  [==========================>>] 100%  ELAPSED TIME: 4.70 s
--------------------------------------------------------------------------------
Loading data to table default.ids_internal partition (datestamp=2016-05-30)
Partition default.ids_internal{datestamp=2016-05-30} stats: [numFiles=1, numRows=1,
totalSize=2, rawDataSize=1]
OK
Time taken: 11.108 seconds
hive> INSERT INTO ids_internal PARTITION (datestamp='2016-05-31') values (11);
Query ID = hdfs_20160703164158_8a2cb0c5-60ef-4212-832b-6cc933d31adf
Total jobs = 1
Launching Job 1 out of 1

Status: Running (Executing on YARN cluster with App id application_1467537169806_0011)

--------------------------------------------------------------------------------
        VERTICES      STATUS   TOTAL   COMPLETED   RUNNING   PENDING   FAILED   KILLED
--------------------------------------------------------------------------------
Map 1 ..........  SUCCEEDED     1         1          0         0         0        0
--------------------------------------------------------------------------------
VERTICES: 01/01  [==========================>>] 100%  ELAPSED TIME: 0.20 s
--------------------------------------------------------------------------------
Loading data to table default.ids_internal partition (datestamp=2016-05-31)
Partition default.ids_internal{datestamp=2016-05-31} stats: [numFiles=1, numRows=1,
totalSize=3, rawDataSize=2]
OK
Time taken: 5.683 seconds
hive> SHOW PARTITIONS ids_internal;
OK
datestamp=2016-05-30
datestamp=2016-05-31
Time taken: 3.684 seconds, Fetched: 2 row(s)
```

We will now create a new subdirectory under this table's directory and add a file to it.

```
hadoop fs -mkdir /apps/hive/warehouse/ids_internal/datestamp=2016-05-21
hadoop fs -put /tmp/2016-05-21.txt /apps/hive/warehouse/ids_internal/datestamp=2016-05-21
```

We can now run the MSCK REPAIR TABLE command to add this new partition to the table:

```
hive> MSCK REPAIR TABLE ids_internal;
OK
Partitions not in metastore:     ids_internal:datestamp=2016-05-21
Repair: Added partition to metastore ids_internal:datestamp=2016-05-21
Time taken: 1.821 seconds, Fetched: 2 row(s)
hive> SHOW PARTITIONS ids_internal;
OK
datestamp=2016-05-21
datestamp=2016-05-30
datestamp=2016-05-31
Time taken: 5.869 seconds, Fetched: 3 row(s)
```

The MSCK repair command checked the subdirectories under /apps/hive/warehouse/ids_internal for the ids_internal table and, because it found a new subdirectory called datestamp=2016-05-21, it added it as a new partition to the ids_internal table. This is particularly useful when you have added many new partition directories and want to update their table definitions all at once. Note that this is valid only with internal tables.

## Rename Partition

You can even rename a table's partition using the ALTER TABLE command. Let's rename the partition that we created in the previous example.

```
hive> ALTER TABLE ids PARTITION (datestamp='2016-05-31') RENAME to PARTITION
(datestamp='31-05-2016');
OK
Time taken: 1.155 seconds
hive> SHOW PARTITIONS ids;
OK
datestamp=2016-05-30
datestamp=31-05-2016
Time taken: 0.679 seconds, Fetched: 2 row(s)
```

```
The ALTER TABLE command in this case, is just updated the partition name in the Hive
Metastore.
```

This command can be used only to modify the external table partitions. You will get the following error if you try to rename the partitions of internal tables.

```
FAILED: Execution Error, return code 1 from org.apache.hadoop.hive.ql.exec.DDLTask. Unable
to rename partition. table new location hdfs://sandbox.hortonworks.com:8020/apps/hive/
warehouse/retail.db/transactions/store=oakdrive is on a different file system than the old
location hdfs://sandbox.hortonworks.com:8020/apps/hive/warehouse/retail.db/transactions/
store=oakwood. This operation is not supported
```

# Modifying Columns

You can modify various columns using the ALTER TABLE command. Let's look at a few options.

## Adding Columns

As the data in the Big Data world grows, one of the key requirements of a schema-on-read architecture is perhaps the ability to modify the schema or table metadata. This flexibility allows users to define various types of metadata on the table and modify it without worrying about modifying the underlying data (only for external tables). You can ALTER a table to add new columns to it using the ALTER TABLE command.

```
hive> ALTER TABLE RETAIL.TRANSACTIONS ADD COLUMNS (loyalty_card boolean);
OK
Time taken: 0.278 seconds
```

The new columns are added to the end of current columns but before the partition columns. The value of the partition column comes from the partition definition and is not stored in the data file itself or mentioned in the column list in CREATE TABLE. Therefore, the partition column is always at the end of the column list when you do SELECT *, although in reality it isn't embedded in the data itself.

You can also replace an entire list of columns in a table using the ALTER TABLE REPLACE COLUMNS command. However, it's better to drop and recreate the table in such a case, because you can then store the new definitions in source control.

# Dropping Tables/Partitions
## Drop Tables

You can drop tables in Hive using the DROP TABLE command. When you run DROP TABLE, the metadata of the table is always deleted. However, Hive only deletes the table data for managed tables. If you have enabled the HDFS trash feature, the data files for the table are moved to the /user/$USER/.trash folder. You can enable this feature by setting the fs.trash.interval parameter in /etc/hadoop/conf/core-site.xml.

```
DROP TABLE <TABLE_NAME>;
```

If you want to drop it from the trash as well, you can include the PURGE keyword as follows:

```
DROP TABLE <TABLE_NAME> PURGE;
```

## Dropping Partitions

You can also drop a partition in Hive using the ALTER TABLE DROP PARTITION command. The command deletes the partition metadata from the Hive Metastore. Just like the DROP TABLE command, Hive deletes the actual partition data only if the table is a managed table. Here is an example of dropping a partition.

```
hive> ALTER TABLE transactions DROP PARTITION (store='oakdrive');
Dropped the partition store=oakdrive
OK
Time taken: 1.105 seconds
```

In this example, the data still exists in HDFS (assuming you used an external table) but queries against the transactions table no longer read from that partition. Therefore, no rows in the result set will have `store=oakdrive` because that partition no longer exists in the table.

## Protecting Tables/Partitions

You can prevent users from dropping tables in Hive by using the `ALTER TABLE ENABLE NO_DROP` command. In a production environment, users typically do not have the privileges to drop the tables. However, this is particularly useful in an environment where the user requires such privileges but you want to protect a particular table from being dropped.

Here is an example of how to alter a table in Hive to prevent it from being deleted:

```
hive> ALTER TABLE transactions ENABLE NO_DROP;
OK
Time taken: 0.239 seconds
hive> DROP TABLE transactions;
FAILED: Execution Error, return code 1 from org.apache.hadoop.hive.ql.exec.DDLTask. Table
transactions is protected from being dropped
```

You can even prevent a table data from being queried by offlining it. This does not prevent another table from accessing the same underlying data.

```
hive> ALTER TABLE transactions ENABLE OFFLINE;
OK
Time taken: 0.285 seconds
hive> SELECT * FROM TRANSACTIONS;
FAILED: SemanticException [Error 10113]: Query against an offline table or partition Table
TRANSACTIONS
```

You can run these two commands at the partition level by specifying the partition name as follows.

```
ALTER TABLE <TABLE_NAME> PARTITION <PARTITION_SPEC> ENABLE OFFLINE;
```

## Other Create Table Command Options

### Create Table as Select (CTAS)

You can also create an internal table using the result set and schema of the output of a query using the `CREATE TABLE AS SELECT` command.

```
hive> CREATE TABLE retail.transactions_top100 AS SELECT * FROM retail.transactions WHERE
custid<101;
```

You can use this feature to extract a subset of a table and store it in another format in a new table. Here is another example that specifies a new format for the target table.

```
hive> CREATE TABLE retail.transactions _top100 STORED AS ORCFILE
    > AS
    > SELECT * FROM retail.transactions WHERE custid<101;
```

Hive has some restrictions on the format of the target table in the CTAS command. The new target table cannot be an external, partitioned, or bucketed table.

## Create Table Like

If you want to copy the schema of an existing table without copying its data, you can use the CREATE TABLE
LIKE command.

```
hive> CREATE TABLE transactions_test LIKE transactions;
OK
Time taken: 0.291 seconds
```

# Data Manipulation Language (DML)

The Hive data manipulation language is the base for all data processing in the Hive ecosystem.

The objectives of this chapter are to:

- Understand the fundamental building blocks of the Hive DML
- Understand the impact of key optional setting
- Combine the fundamental building blocks to achieve data processing

---

To achieve the maximum learning experience, you should complete the chapter's examples in the order they are presented, as later examples use previous data structures. This is a much more structured chapter than the others in this book to efficiently explain the syntax of each DML topic.

---

## Loading Data into Tables

Processing data into information requires data to be present. The Hive environment will accept any data that can be structured in a delimited format.

Data is loaded into the platform using the following DML process.

To load data into the platform you need two components:

- Data to load from (a source)
- A table to load the data into (a target)

---

There is no transformation while loading data into tables, as Hive only performs a move/copy of the data ready for system to use.

---

© Scott Shaw, Andreas François Vermeulen, Ankur Gupta, David Kjerrumgaard 2016
S. Shaw et al., *Practical Hive*, DOI 10.1007/978-1-4842-0271-5_5

# Loading Data Using Files Stored on the Hadoop Distributed File System

Hive supports uploading files from the Hadoop Distributed File System (HDFS). This is the most fundamental method of moving data into the Hive ecosystem.

The Hive syntax is as follows:

```
LOAD DATA [LOCAL] INPATH 'filepath' [OVERWRITE] INTO TABLE tablename
```

Here is the syntax explained:

| | |
|---|---|
| LOAD DATA | Keywords for loading data in Hive. |
| LOCAL | If included, enables the users to load data from their local files. If omitted, the files are loaded from the path set in the Hadoop configuration variable fs.default.name. |
| INPATH 'filepath' | If LOCAL is used: file:///user/hive/example If LOCAL is omitted: hdfs://namenode:9000/user/hive/example |
| OVERWRITE | If included, enables the users to load data into an already populated table and replace the previous data. If omitted, enables the users to load data into an already populated table and append the new data to previous data. |
| INTO TABLE tablename | The tablename is the name of a table that exists in Hive. Use CREATE TABLE tablename. |

## Using Hive to Upload a Data File

The following Hive commands enable you to upload a data file called Person001.csv into a table called census.person.

---

The data sets can be downloaded from www.apress.com/9781484202722.

---

For the purposes of this chapter, you need to use:

```
$HIVE_HOME/bin/hive
```

This example uses the example script called Script_PersonTable.txt.

The Hive script to use is:

```
## Create a new database
CREATE DATABASE census;

## Use the new database
USE census;
```

```
## Create a new table
CREATE TABLE person (
  persid          int,
  lastname        string,
  firstname       string
)
ROW FORMAT  DELIMITED  FIELDS TERMINATED BY ',';

## Load data into the new table from csv file
LOAD DATA LOCAL INPATH 'file:///root/hive/example/person001' OVERWRITE INTO TABLE person;

## Check if the data is in table
SELECT persid, lastname, firstname
FROM person;
```

The following will show if you use the script on the Hive command line:

```
hive> CREATE DATABASE census;
OK
Time taken: 1.486 seconds

hive> USE census;
OK
Time taken: 0.66 seconds

hive> CREATE TABLE person (
    >    persid          int,
    >    lastname        string,
    >    firstname       string
    >)
    >ROW FORMAT  DELIMITED  FIELDS TERMINATED BY ',';
OK
Time taken: 3.28 seconds

hive> LOAD DATA LOCAL INPATH 'file:///root/hive/example/person001' OVERWRITE INTO TABLE
person;
Loading data to census.person
Table census.person stats: (numFiles=1, numRows=0, totalSize=1265, rawDataSize=0)
OK
Time taken: 4.393 seconds
```

Test if all the data is loaded; the results should be 80 records (we show the first 10 records):

```
hive> SELECT persid, lastname, firstname FROM person;
OK
2       SMITH           AARON
3       SMITH           ABDUL
4       SMITH           ABE
5       SMITH           ABEL
6       SMITH           ABRAHAM
7       SMITH           ABRAM
```

```
8       SMITH           ADALBERTO
9       SMITH           ADAM
10      SMITH           ADAN
11      JOHNSON         AARON
..
..
Time taken: 4.241 seconds, Fetched: 80 record(s)
```

## Loading Data Using Queries

Hive supports loading data queried from existing tables into the Hive ecosystem.

The Hive syntax is as follows:

```
INSERT [OVERWRITE]
TABLE tablename1 [IF NOT EXISTS]
SELECT select_fields FROM from_statement;
```

Here is the syntax explained:

| | |
|---|---|
| INSERT | Keywords for loading data into a Hive table. |
| OVERWRITE | If included, enables the users to load data into an already populated table and replace the previous data.<br>If omitted, enables the users to load data into an already populated table and append the new data to previous data. |
| TABLE tablename | The tablename is the name of a table that exists in Hive.<br>Use CREATE TABLE tablename. |
| IF NOT EXISTS | If the IF NOT EXISTS is included in the command, the Hive command will create a table in the current database.<br>If omitted, it will fail if the table does not exist. |
| SELECT<br>  select_fields<br>FROM<br>  from_statement | This can be any SELECT command against the Hive ecosystem. |

## Using an Existing Table to Create a New Table

This exercise enables you to upload a data query from a table called census.person into a table called census.personhub.

The example uses the example script Script_PersonHub.txt.

The complete script is:

```
## Use existing database
USE census;
```

```
## Create new table
CREATE TABLE personhub (
  persid          int

);

## Insert data into table, overwriting existing data in table
INSERT OVERWRITE
TABLE personhub
SELECT DISTINCT personId FROM Person;

## Check if data in table
SELECT
  persid
FROM
  personhub;
```

The following will show if you use the script on the Hive command line:

```
hive> USE census;
OK
Time taken: 0.664 seconds

hive> CREATE TABLE personhub ( persid int );
OK
Time taken: 3.098 seconds

hive> INSERT OVERWRITE TABLE personhub SELECT DISTINCT personId FROM Person; );
Query ID = root_201606081616_9defdc9d-5d2d-46aa-87e1-a7e7247b2362
Total jobs = 1
Launching Job 1 out of 1

Status: Running (Executing on YARN cluster with App id application_1441527339718_004
-----------------------------------------------------------------------
VERTICES        STATUS  TOTAL COMPLETED  RUNNING  PENDING  FAILED  KILLED
-----------------------------------------------------------------------
MAP 1 ........ SUCCEEDED   1       1        0        0        0       0
Reducer 2 .... SUCCEEDED   1       1        0        0        0       0
-----------------------------------------------------------------------
VERTICES: 02/02 [======================>>] 100% ELAPSED TIME: 31.84 s
-----------------------------------------------------------------------
Loading data to table census.personhub
Table census.personhub stats: [numFiles=1, numRows=80, totalSize=232, rawDataSize=152]
OK
Time taken: 39.003 seconds
```

The results should be 80 records (we show the first five):

```
hive> SELECT persid FROM personhub;
OK
2
3
4
5
6
..
.. ( Only shown 5 record - 75 records removed ...)
Time taken: 2.7.64 seconds, Fetched: 80 record(s)
```

Now we upload the data again to test the removal of the OVERWRITE parameter.

```
USE census;
```

```
INSERT OVERWRITE TABLE personhub SELECT DISTINCT persid FROM Person;
```

Test if all the data is loaded without removing the previous data:

```
SELECT persid FROM personhub;
```

The results should be 160 records (only five are shown):

```
hive> USE census;
OK
Time taken: 0.662 seconds

hive> INSERT OVERWRITE TABLE personhub SELECT DISTINCT personId + 1000 FROM Person; );
Query ID = root_201606081622_8defde9d-5d2d-46aa-87e1-a9e7247b2362
Total jobs = 1
Launching Job 1 out of 1

Status: Running (Executing on YARN cluster with App id application_1441527339718_005
--------------------------------------------------------------------------
VERTICES        STATUS   TOTAL  COMPLETED  RUNNING  PENDING  FAILED  KILLED
--------------------------------------------------------------------------
MAP 1 ........ SUCCEEDED   1        1         0        0        0       0
Reducer 2 .... SUCCEEDED   1        1         0        0        0       0
--------------------------------------------------------------------------
VERTICES: 02/02 [======================>>] 100% ELAPSED TIME: 31.84 s
--------------------------------------------------------------------------
Loading data to table census.personhub
Table census.personhub stats: [numFiles=1, numRows=80, totalSize=232, rawDataSize=152]
OK
Time taken: 41.411 seconds
```

```
hive> SELECT persid FROM personhub;
OK
2
3
4
1002
1003
..
..
Time taken: 2.7.64 seconds, Fetched: 160 record(s)
```

## Writing Data into the File System from Queries

Hive supports loading data queried back into the Hadoop Distributed File System.

The Hive syntax is as follows:

```
INSERT [OVERWRITE]
DIRECTORY directoryname
SELECT select_fields FROM from_statement;
```

Here is the syntax explained:

| | |
|---|---|
| `INSERT` | Keywords for loading data into a Hive directory. |
| `OVERWRITE` | If included, enables the users to load data into an already populated directory and replace the previous data. If omitted, enables the users to load data into an already populated directory and append the new data to previous data. |
| `DIRECTORY directoryname` | The `directoryname` is the name of a directory that exists in the Hadoop Distributed File System. Use `hadoop fs -mkdir directoryname` to create a directory. |
| `SELECT select_fields FROM from_statement` | This can be any `SELECT` command against the Hive ecosystem. |

## Using an Existing Table to Create an Output Directory

This exercise enables you to upload a data query from a table called `person` into an output directory.
The example use the example script `Script_PersonDirectory.txt`:

The complete script is:

```
hadoop fs -mkdir 'exampleoutput'
hive

USE census;
```

```
INSERT OVERWRITE DIRECTORY 'exampleoutput'
ROW FORMAT DELIMITED FIELDS TERMINATED BY ','
SELECT persid, firstname, lastname
FROM person;

exit;
```

Test if all the data is loaded:

```
hadoop fs -cat 'exampleoutput/000000_0'
```

The following shows if you use the script on the Hive command line:

```
hive> INSERT OVERWRITE DIRECTORY 'exampleoutput'
    > ROW FORMAT DELIMITED FIELDS TERMINATED BY ','
    > SELECT persid, firstname, lastname FROM person;

Query ID = root_201606081622_8dedde9d-9d2d-46ab-89e1-a9e7249b2362
Total jobs = 1
Launching Job 1 out of 1

Status: Running (Executing on YARN cluster with App id application_1441527339718_012
--------------------------------------------------------------------------------
VERTICES        STATUS    TOTAL COMPLETED  RUNNING  PENDING  FAILED  KILLED
--------------------------------------------------------------------------------
MAP 1 ........ SUCCEEDED    1       1         0        0        0       0
--------------------------------------------------------------------------------
VERTICES: 01/01 [======================>>] 100% ELAPSED TIME: 22.05 s
--------------------------------------------------------------------------------
Loading data to table census.personhub
Table census.personhub stats: [numFiles=1, numRows=80, totalSize=232, rawDataSize=152]
OK
Time taken: 66.685 seconds

hive> exit;

> hadoop fs -cat 'exampleoutput/000000_0'

2       SMITH         AARON
3       SMITH         ABDUL
4       SMITH         ABE
5       SMITH         ABEL
6       SMITH         ABRAHAM
7       SMITH         ABRAM
8       SMITH         ADALBERTO
9       SMITH         ADAM
10      SMITH         ADAN
11      JOHNSON       AARON
..
..
```

# Inserting Values Directly into Tables

Hive supports loading data directly into tables using a series of static values.

The Hive syntax is as follows:

```
INSERT
INTO TABLE tablename
VALUES
(row_values1),
(row_values2);
```

Here is the syntax explained:

| | |
|---|---|
| INSERT | Keywords for loading data into a Hive directory. |
| TABLE tablename | The `tablename` is the name of a table that exists in Hive. Use `CREATE TABLE` tablename. |
| VALUES (row_values1), (row_values2) | The `row_values1` and `row_values2` values are individual records of same format other than the table's records. |

## Adding Extra Records to an Existing Table

This exercise enables you to upload a record directly into a table called `personhub`.
The example uses the example script `Script_PersonValues.txt`.

The complete script is:

```
USE census;

INSERT
INTO TABLE personhub
VALUES
(0);
```

Test if all the data is loaded:

```
USE census;

SELECT persid
FROM personhub
WHERE persid = 0;
```

The following shows if you use the script on the Hive command line:

```
hive> USE census;
OK
Time taken: 0.662 seconds
```

```
hive> INSERT INTO TABLE personhub VALUES (0);
Query ID = root_201606081622_8defde5d-5d2d-46aa-89e1-a9e7247b2362
Total jobs = 1
Launching Job 1 out of 1

Status: Running (Executing on YARN cluster with App id application_1441527339718_015
-----------------------------------------------------------------------
VERTICES        STATUS    TOTAL COMPLETED  RUNNING  PENDING  FAILED  KILLED
-----------------------------------------------------------------------
MAP 1 ........ SUCCEEDED     1       1        0        0       0       0
-----------------------------------------------------------------------
VERTICES: 02/02 [======================>>] 100% ELAPSED TIME: 51.05 s
-----------------------------------------------------------------------
Loading data to table census.personhub
Table census.personhub stats: [numFiles=1, numRows=80, totalSize=232, rawDataSize=152]
OK
Time taken: 41.411 seconds
```

The results should be a single record:

```
hive> SELECT persid FROM personhub WHERE persid = 0;
OK
0
Time taken: 5.493 seconds, Fetched: 1 record(s)
```

## Updating Data Directly in Tables

Hive supports updating data directly into tables.

The Hive syntax is as follows:

```
UPDATE tablename
SET column = value
[WHERE expression];
```

Here is the syntax explained:

| | |
|---|---|
| UPDATE | Keywords for updating values in a table. |
| tablename | The tablename is the name of a table that exists in Hive. Use CREATE TABLE tablename. |
| SET column = value | The SET command updates the column with a value. |
| [WHERE expression] | WHERE can be used to pick specific column values for a change query. |

# Updating Records in an Existing Table

This exercise enables you to update data directly in a table called person20.

The example uses the script Script_PersonUpdate.txt.

The complete script is:

```
USE census;

CREATE TABLE census.person20 (
  persid          int,
  lastname        string,
  firstname       string
)
CLUSTERED BY (persid) INTO 1 BUCKETS
STORED AS orc
TBLPROPERTIES('transactional' = 'true');

INSERT INTO TABLE person20 VALUES (0,'A','B'),(2,'X','Y');
```

Test if the data is updated:

```
SELECT *
FROM
  census.person20;
```

The results should be two records.

```
OK
0       A       B
2       X       Y
```

Now perform the update:

```
USE census;

UPDATE
  census.person20
SET lastname = 'SS'
WHERE
 persid = 0;

SELECT *
FROM
  census.person20;
```

The results should two records.

```
OK
0       SS      B
2       X       Y
```

## Deleting Data Directly in Tables

Hive supports deleting data directly in tables.

The Hive syntax is as follows:

```
DELETE tablename
[WHERE expression];
```

Here is the syntax explained:

| | |
|---|---|
| DELETE | Keywords for deleting values in a table. |
| tablename | The tablename is the name of a table that exists in Hive. Use CREATE TABLE tablename. |
| [WHERE expression] | WHERE can be used to pick specific column values to delete the query. |

## Updating Records in an Existing Table

This exercise enables you to update records directly in a table called person30.
The example uses the script Script_PersonDelete.txt.

The complete script is:

```
USE census;

CREATE TABLE census.person30 (
   persid       int,
   lastname     string,
   firstname    string
)
CLUSTERED BY (persid) INTO 1 BUCKETS
STORED AS orc
TBLPROPERTIES('transactional' = 'true');

INSERT INTO TABLE census.person30
VALUES (0,'A','B'),(2,'X','Y');

SELECT *
FROM census.person30;
```

The results should be two records.

```
OK
0       A       B
2       X       Y
```

To delete a record:

```
USE census;

DELETE FROM census.person30
WHERE persid = 0;

SELECT *
FROM census.person30;
```

The results should be one record.

```
OK
0       A       B
2       X       Y
```

## Creating a Table with the Same Structure

Hive supports creating a new table from an existing table's structure.

The Hive syntax is as follows:

```
CREATE
TABLE blank_tablename
LIKE tablename;
```

Here is the syntax explained:

| | |
|---|---|
| CREATE TABLE | Keywords for creating a table. |
| Blank_tablename | The tablename is the name of a table that's created. |
| LIKE | Keyword to ensure the same structure is used. |
| tablename | The tablename is the name of a table that exists in Hive. Use CREATE TABLE tablename. |

## Using an Existing Table to Create a New Table with the Same Structure

This exercise enables you to create a table called personhub2 using the structure of a table called personhub. The example uses the script Script_PersonLike.txt.

The complete script is:

```
USE census;

CREATE TABLE person40 LIKE person;

SELECT * FROM person40;
```

Test if the data is updated:

```
INSERT INTO TABLE person40 VALUES (0,'Bob','Burger'),(1,'Charlie','Clown');
SELECT * FROM person40;
```

The results should be two records.

```
OK
0        A        B
2        X        Y
```

# Joins

## Using Equality Joins to Combine Tables

Hive supports equality joins between tables to enable you to combine data from two tables.

The Hive syntax is as follows:

```
SELECT table_fields
FROM table_one
JOIN table_two
ON (table_one.key_one = table_two.key_one
AND table_one.key_two = table_two.key_two);
```

Here is the syntax explained:

| | |
|---|---|
| `SELECT table_fields` | Keywords to select of a range of fields from both tables. |
| `FROM table_one`<br>`JOIN table_two` | Lists the two tables that are joined to retrieve the `table_fields`. |
| `ON (table_one.key_one = table_two.key_one`<br>`AND table_one.key_two = table_two.key_two)` | Lists the equality rules to join the two tables. |

## Joining Tables in Hive

This exercise enables you to create a join between two tables called `census.personname` and `census.address`.

The example uses the script `Script_EqualJoin.txt`.

The complete script is:

```
USE census;
CREATE TABLE census.personname (
  persid         int,
  firstname      string,
  lastname       string
)
```

```
CLUSTERED BY (persid) INTO 1 BUCKETS
STORED AS orc
TBLPROPERTIES('transactional' = 'true');

INSERT INTO TABLE census.personname
VALUES
(0,'Albert','Ape'),
(1,'Bob','Burger'),
(2,'Charlie','Clown'),
(3,'Danny','Drywer');

CREATE TABLE census.address (
    persid          int,
    postname        string
)
CLUSTERED BY (persid) INTO 1 BUCKETS
STORED AS orc
TBLPROPERTIES('transactional' = 'true');
INSERT INTO TABLE census.address
VALUES
(1,'KA13'),
(2,'KA9'),
(10,'SW1');
```

You now have two tables called census.personname and census.address. Now you perform the join:

```
SELECT personname.firstname,
    personname.lastname,
    address.postname
FROM
    census.personname
JOIN
    census.address
ON (personname.persid = address.persid);
```

The results of the join are as follows:

```
OK
Bob       Burger       KA13
Charlie   Clown        KA9
```

## Using Outer Joins

Hive supports equality joins between tables using LEFT, RIGHT, and FULL OUTER joins, where keys have no match.

The Hive syntax is as follows:

```
SELECT table_fields
FROM table_one
[LEFT, RIGHT, FULL OUTER] JOIN table_two
ON (table_one.key_one = table_two.key_one
AND table_one.key_two = table_two.key_two);
```

Here is the syntax explained:

| | |
|---|---|
| `SELECT table_fields` | Keywords to select a range of fields from both tables. |
| `FROM table_one`<br>`LEFT JOIN table_two` | Lists the two tables that are joined to retrieve the `table_fields`. The LEFT join will result in including fields values from `table_one` that match the `where` statement and fields values from `table_two` that match and don't match the `where` statement. |
| `FROM table_one`<br>`RIGHT JOIN table_two` | Lists the two tables that are joined to retrieve the `table_fields`. The RIGHT join will result in including fields values from `table_one` that match the `where` statement and fields values from `table_two` that match and don't match the `where` statement. |
| `FROM table_one`<br>`FULL OUTER JOIN table_two` | Lists the two tables that are joined to retrieve the `table_fields`. The FULL OUTER join will result in including fields values from `table_two` that match and don't match the `where` statement and fields values from `table_two` that don't match the `where` statement. |
| `ON (table_one.key_one = table_two.key_one`<br>`AND table_one.key_two = table_two.key_two)` | Lists the equality rules to join the two tables. |

## Joining Tables in Hive Using Left Join

Hive supports equality joins between tables to enable you to combine data from two tables.
The example uses the script `Script_OuterJoin.txt`.

The complete script is:

```
USE census;

SELECT personname.firstname,
  personname.lastname,
  address.postname
FROM
  census.personname
LEFT JOIN
  census.address
ON (personname.persid = address.persid);
```

The results should be four records.

```
OK
Albert    Ape           NULL
Bob       Burger        KA13
Charlie   Clown         KA9
Danny     Drywer        NULL
```

## Joining Tables in Hive Using Right Join

Let's do a right join:

```
SELECT personname.firstname,
  personname.lastname,
  address.postname
FROM
  census.personname
RIGHT JOIN
  census.address
ON (personname.persid = address.persid);
```

The results should be three records.
Here are the results of the right join:

```
OK
Bob       Burger        KA13
Charlie   Clown         KA9
NULL      NULL          SW1
```

## Joining Tables in Hive Using a Full Outer Join

Now for an outer join:

```
SELECT personname.firstname,
  personname.lastname,
  address.postname
FROM
  census.personname
FULL OUTER JOIN
  census.address
ON (personname.persid = address.persid);
```

The results should be five records.

```
OK
Albert    Ape           NULL
Bob       Burger        KA13
Charlie   Clown         KA9
Danny     Drywer        NULL
NULL      NULL          SW1
```

93

## Using Left Semi-Joins

Hive supports nested joins between tables. Consider a nested join like the following:

```
SELECT a.key, a.value
FROM a
WHERE a.key in
 (SELECT b.key
  FROM B);
```

> This query will not work in Hive due to the distributed processing.
> Hive can handle the query and uses a SEMI JOIN command.
> The Hive syntax is as follows:

```
SELECT table_fields
FROM table_one
LEFT SEMI JOIN table_two
ON (table_one.key_one = table_two.key_one);
```

Here is the syntax explained:

| | |
|---|---|
| `SELECT table_fields` | Keywords to select a range of fields from both tables. |
| `FROM table_one LEFT SEMI JOIN table_two` | Lists the two tables that are semi-joined to retrieve the `table_fields`. |
| `ON (table_one.key_one = table_two.key_one);` | Lists the equality rules to join the two tables. |

## Performing a Semi-Join

Hive supports semi-joins between tables to enable you to combine data from two tables.
The example uses the script Script_SemiJoin.txt.

The complete script is:

```
USE census;

SELECT
  personname.firstname,
  personname.lastname
FROM
  census.personname
LEFT SEMI JOIN
  census.address
ON (personname.persid = address.persid);
```

The results should be two records.

```
OK
Bob      Burger      KA13
Charlie  Clown       KA9
```

# Using Join with Single MapReduce

Hive supports join using single MapReduce between multiple tables if the common key is used in a chain of joins.

The Hive syntax is as follows:

```
SELECT table_one.key_one, table_two.key_one, table_three.key_one
FROM table_one JOIN table_two
ON (table_one.key_one = table_two.key_one)
JOIN table_three
ON (table_three.key_one = table_two.key_one);
```

Here is the syntax explained:

| | |
|---|---|
| `SELECT table_one.key_one, table_two.key_one, table_three.key_one` | Keywords to select a range of fields from all tables. |
| `FROM table_one JOIN table_two` | Lists the first and second tables that are joined to retrieve the `table_fields`. |
| `ON (table_one.key_one = table_two.key_one)` | Lists the equality rules to join the first and second tables. |
| `JOIN table_three` | Lists the third table that is joined to retrieve the `table_fields`. |
| `ON (table_three.key_one = table_two.key_one)` | Lists the equality rules to join the third table. |

## Joining Three Tables in One MapReduce

This exercise enables you to join three tables in one MapReduce.
The example uses the script `Script_MultiJoin.txt`.

The complete script is:

```
USE census;

CREATE TABLE census.account (
  persid      int,
  bamount     int
)
CLUSTERED BY (persid) INTO 1 BUCKETS
STORED AS orc
TBLPROPERTIES('transactional' = 'true');
INSERT INTO TABLE census.account
VALUES
(1,12),
(2,9);
```

```
SELECT
  personname.firstname,
  personname.lastname,
  address.postname,
  account.bamount
FROM
  census.personname
JOIN
  census.address
ON (personname.persid = address.persid)
JOIN
  census.account
ON (personname.persid = account.persid);
```

The results should be two records.

```
OK
Bob       Burger      KA13    12
Charlie   Clown       KA9     9
```

## Using Largest Table Last

Hive performs joins by buffering the first tables of the join and then mapping the last table against them.

It's good practice to always list the biggest table last because this speeds up the processing.

The Hive syntax one is as follows:

```
SELECT table_one.key_one, table_two.key_one, table_three.key_one
FROM table_one JOIN table_two
ON (table_one.key_one = table_two.key_one)
JOIN table_three
ON (table_three.key_one = table_two.key_one);
```

Here is syntax one explained:

| | |
|---|---|
| table_one and table_two | Buffered in memory. |
| table_three | Mapped directly from disk. |

The Hive syntax two is as follows:

```
SELECT table_one.key_one, table_two.key_one, table_three.key_one
FROM table_one JOIN table_three
ON (table_one.key_one = table_three.key_one)
JOIN table_two
ON (table_two.key_one = table_three.key_one);
```

Here is syntax two explained:

| | | |
|---|---|---|
| `table_one` and | `table_three` | Buffered in memory. |
| `table_two` | | Mapped directly from disk. |

## Transactions

Hive supports ACID-compliant transactions. This enables the support of transactions that are confirmed to completion by ensuring data integrity in the Hive database.

This is not a default setting for most Hive installations, as it will have a performance impact due to the extra processing required to ensure ACID compliance.

## What Is ACID and Why Use It?

ACID stands for four traits of database transactions:

- Atomicity—An operation either succeeds completely or fails; operations do not leave incomplete data in the system.

- Consistency —Once an operation completes, the results of that operation are visible to every subsequent operation.

- Isolation—Operations completed by one user do not cause unexpected side effects for other users.

- Durability—Once an operation is complete, it will be preserved even if the machine or system experiences a failure.

These behaviors are mandatory to ensure transaction functionality.

If your operations are ACID compliant, the system will ensure your processing is protected against any failures.

## Hive Configuration

Hive supports transactions by setting the correct parameters.

To enable transactions, the following configurations need to be set. These configuration parameters must be set appropriately to turn on transaction support in Hive:

- `hive.support.concurrency - true`

- `hive.enforce.bucketing - true`

- `hive.exec.dynamic.partition.mode - nonstrict`

- `hive.txn.manager - org.apache.hadoop.hive.ql.lockmgr.DbTxnManager`

- `hive.compactor.initiator.on` – true on one instance of the Thrift metastore service

- `hive.compactor.worker.threads` – 10 for an instance of the Thrift metastore service

Use this specific table format:

```
CREATE TABLE table_one (
  keyField           int,
  valueFieldOne      string,
  valueFieldTwo      string
)
CLUSTERED BY (keyField) INTO x BUCKETS
STORED AS orc
TBLPROPERTIES('transactional' = 'true');
```

# CHAPTER 6

∎ ∎ ∎

# Loading Data into Hive

Let's say you have built a data lake in your organization and one of the lines of business has requested for a new use case to be implemented, for example, a 360 view of the customer. When you consider the details of the use case, you find that analytics needs to occur on all the customer data residing in the existing operational systems, data warehouse, and on all new data getting generated from social media, customer service, and call centers, to get a complete picture of the customer. Hadoop, being a general-purpose, large-scale distributed processing platform, is quite suitable for this.

However, before you can run any kind of analytics in this data lake, the first task is to load the data. Historically, it was a common pattern to extract data from operational systems and load it in a data warehouse in batch form. But for this particular use case, you will need to load structured data from relational database, tweets from Twitter, feeds from Facebook, and audio call records from the call center system.

Previously, there wasn't a single tool in the Hadoop ecosystem that was suitable to load data from all systems and in all formats. Instead, the community wrote a variety of tools that work best with some systems and were suitable for loading data in specific formats. As you can imagine, loading data using various tools in different formats from various systems can soon become a complex problem. The complexity of loading data can further be impacted by a few other factors. The frequency in which the data is loaded from a source system might also have an impact on the best tool. By way of interest, Apache Nifi, a part of the Hortonworks Data Flow Platform, has become an example of such a comprehensive tool, for all types of data loading and ingestion scenarios.

Regardless of the type of the source, the structure of the data, and the tools used to load it, all data in a Hadoop-based platform gets stored in HDFS. Since Hive is a SQL layer on Hadoop, all data needs to be loaded into HDFS before it becomes available for querying through Hive.

In this chapter we look at the common tools that can be used to load various types of data into HDFS. Some of the tools require the manual addition of Hive Metadata, whereas other tools automatically update Hive Metastore to make the newly added data available for analysis through Hive.

## Design Considerations Before Loading Data

Before you start to populate any data in Hadoop, here are some key aspects that you should consider:

- It is imperative to design the filesystem layout of HDFS to store various types of data. This will ensure easier management, discovery, and access control of the data to various users.

- If you are loading structured data from a relational database, you will need to decide whether to create a similar schema in Hive or a different data model.

- The format in which the data is stored in HDFS (such as ORCfile, RCfile, AVRO, Parquet, and so on) can impact the performance of the queries run through Hive. Most of the recent performance optimizations in Hive only work with ORCFile; we will see more details of this in Chapter 9, "Performance Tuning: Hive".

© Scott Shaw, Andreas François Vermeulen, Ankur Gupta, David Kjerrumgaard 2016
S. Shaw et al., *Practical Hive*, DOI 10.1007/978-1-4842-0271-5_6

- Depending on the volume and access pattern of the data, you should also decide the most suitable compression algorithm—for example, Snappy, Zlib, LZO, etc.—to apply when the data is copied to HDFS.

- It is recommended not to store a large number of very tiny files in HDFS. This leads to an inefficient namespace usage of NameNode. Hence, it is important that you decide the appropriate file size and configuration for all files in HDFS.

- The loading patterns of the data can be a one-time batch, frequent batches, or real-time ingestion. The choice of tools used to load the data can be driven by the loading patterns.

# Loading Data into HDFS

This section describes techniques and tools for moving data into Hadoop. There are a variety of ways to get data into Hadoop, from simple Hadoop shell commands to more sophisticated processes. We discuss these processes and also look at few examples. These methods assume that you have privileges on the HDFS directory into which you are copying the files.

## Ambari Files View

Ambari Files View is one of the views shipped with Ambari. The view provides a web user interface for browsing HDFS, creating/removing directories, downloading/uploading files, and so on. The cluster must have HDFS and WebHDFS deployed in order to use the Ambari Files View.

You can upload a file to HDFS using Ambari Files View as follows:

1. Log in to Ambari.

2. Open the Ambari Files View by hovering the mouse over the Your Views menu to the left of the login username, in order to view a drop-down list of all available view instances (as shown in Figure 6-1).

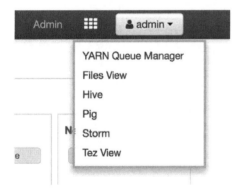

***Figure 6-1.*** *List of Ambari views*

3. Click on Files View to browse the HDFS filesystem (as shown in Figure 6-2). The actual name of the Files View instance might be different in your cluster.

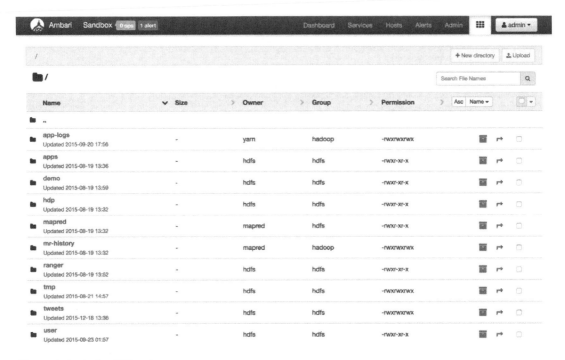

***Figure 6-2.*** *Ambari Files view*

4.  Select the HDFS directory where you would like to upload the file.

5.  Click on Upload and Browse to open the File Browse window (as shown in Figure 6-3).

***Figure 6-3.*** *Browse local files*

6.  Select the file that you want to upload and click on Upload.

7.  The uploaded file should now be visible in the list of files listed in the current directory (as shown in Figure 6-4).

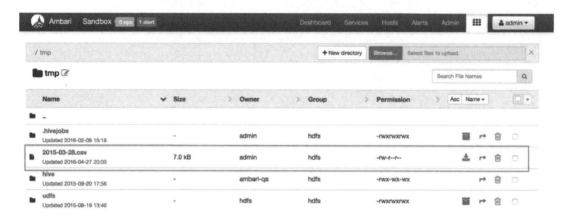

**Figure 6-4.**  *File uploaded using Ambari Files View*

## Hadoop Command Line

Hadoop has a built-in hadoop command line that you can use to move files from the local filesystem to HDFS. This command-line tool is quite handy when you don't have access to Ambari but have access through the shell. This command-line script has many commands that can be used to perform other operations on HDFS. However, in this section, we will only discuss the options to upload files to HDFS. All the other commands are beyond the scope of this book.

Here is the syntax to copy files in HDFS:

```
hadoop fs –put source_path hdfs_path
```

Let's look at an example to copy another CSV file to the HDFS /tmp directory:

```
[hdfs@sandbox tmp]$ hadoop fs -put /tmp/2014-01-28.csv /tmp/
[hdfs@sandbox tmp]$ hadoop fs -ls /tmp/
Found 6 items
drwxrwxrwx   - admin      hdfs          0 2016-05-01 21:48 /tmp/.hivejobs
-rw-r--r--   1 hdfs       hdfs       3864 2016-06-14 22:14 /tmp/2014-01-28.csv
-rw-r--r--   3 admin      hdfs       7168 2016-04-27 19:03 /tmp/2015-03-28.csv
drwx-wx-wx   - ambari-qa  hdfs          0 2015-09-20 16:56 /tmp/hive
drwxr-xr-x   - root       hdfs          0 2016-05-01 22:24 /tmp/root
drwxrwxrwx   - hdfs       hdfs          0 2015-08-19 12:46 /tmp/udfs
```

## HDFS NFS Gateway

NFS Gateway is a stateless daemon that translates the NFS protocol to HDFS access protocols. It allows the clients to mount HDFS and interact with it through NFS, as if it were a part of their local filesystem. Many instances of such daemon can be run to provide high throughput read/write access to HDFS from multiple clients. Before a client can mount HDFS using NFS Gateway, it must be installed on one of the data nodes or NameNodes of the Hadoop cluster. Once the HDFS filesystem is mounted using NFS Gateway, the user can simply copy files using the OS command line to HDFS.

# Sqoop

As shown in Figure 6-5, Sqoop is used to transfer data between structured data stores such as relational databases, enterprise data warehouses, and NoSQL systems and Hadoop. It extracts data from an external system on to HDFS and can also populate tables in Hive and HBase. Sqoop automates most of this process, relying on the database to describe the schema for the data to be imported.

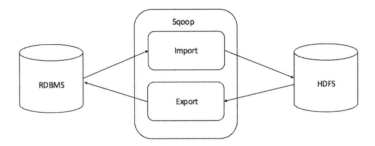

***Figure 6-5.*** *Sqoop's workflow*

Sqoop has a connector-based architecture that connects to various external systems. These connectors use a set of JDBC drivers for communication with various systems. Some of the external systems, which do not provide a JDBC interface, can also be accessed using these connectors. There are different connectors for different external systems. Depending on which external system you can want to connect to from Sqoop, you can add the appropriate plug-in. Some of the common connectors included with Sqoop are MySQL, Netezza, Oracle, PostgreSQL, Microsoft SQL Server, and Teradata.

In this section, we look at the general architecture of Sqoop and study some examples to import data from a MySQL database.

## How Sqoop Works

Sqoop is used for bulk transfers of data. Internally it uses map reduce to read/write data to HDFS. When you run a Sqoop command, the data set that needs to be transferred is divided into various chunks and a map-job is assigned to each data chunk. These data slices are worked in parallel, which is why Sqoop is able to transfer bulk data efficiently.

Figure 6-6 represents a Sqoop import job with a parallelism of four to load data into HDFS.

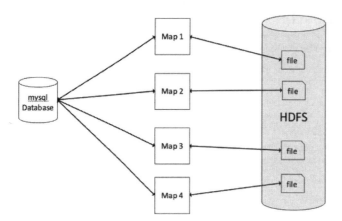

**Figure 6-6.** *Sqoop import architecture*

## Sqoop Examples

Let's look at some examples that move data using Sqoop.

### Importing a Table into HDFS

```
sqoop import --connect jdbc:mysql://localhost/test --table TEST1 --username root --m 1
```

This command will export the table TEST1 from the `test` MySQL database and store it in HDFS in the directory `/user/<user>/TEST1/part-m-00000` file.

### Importing a Table into a Specific Directory in HDFS

```
sqoop import --connect jdbc:mysql://localhost/test --table TEST1 --username root --m 1
--target-dir /hive/tables/TEST1/
```

In this example, the contents of the TEST1 table will be stored in the `/hive/tables/TEST1` directory in HDFS.

### Importing All Tables from a Database to HDFS

```
sqoop import-all-tables --connect jdbc:mysql://localhost/test --username root
```

This command will import all tables in the `test` database into HDFS. The Sqoop import job creates a directory for each table under `/user/root`. We can see the list of imported tables as follows:

```
[root@sandbox ~]# hadoop fs -ls /user/root
Found 5 items
drwx------   - root hdfs          0 2016-04-30 21:18 /user/root/.Trash
drwxr-xr-x   - root hdfs          0 2015-09-20 16:56 /user/root/.hiveJars
drwx------   - root hdfs          0 2016-04-30 22:05 /user/root/.staging
drwxr-xr-x   - root hdfs          0 2016-06-14 22:24 /user/root/TEST1
drwxr-xr-x   - root hdfs          0 2016-06-14 22:24 /user/root/TEST2
```

## Importing a Table into Hive

```
sqoop import --connect jdbc:mysql://localhost/test --table TEST1 --username root  --m 1
--hive-import
```

This command will import the TEST1 table into HDFS but also add its metadata to Hive. We can verify the data in Hive as follows.

```
 hive> use default;
OK
Time taken: 1.453 seconds
hive> select count(*) from test1;
Query ID = root_20160614222847_b86f0300-0a22-49fe-a56f-e997c3e7e0e2
Total jobs = 1
Launching Job 1 out of 1

Status: Running (Executing on YARN cluster with App id application_1465942169140_0009)

--------------------------------------------------------------------------------
        VERTICES      STATUS  TOTAL  COMPLETED  RUNNING  PENDING  FAILED  KILLED
--------------------------------------------------------------------------------
Map 1 ..........   SUCCEEDED     1         1        0        0       0       0
Reducer 2 ......   SUCCEEDED     1         1        0        0       0       0
--------------------------------------------------------------------------------
VERTICES: 02/02  [==========================>>] 100%  ELAPSED TIME: 4.98 s
--------------------------------------------------------------------------------
OK
3145728
Time taken: 13.049 seconds, Fetched: 1 row(s)
```

## Importing a Table into Hive with Data Stored as an ORC Table

```
sqoop import --connect jdbc:mysql://localhost/test --table TEST10 --username root  --m 1
--hcatalog-database default --hcatalog-table TEST10_ORC --create-hcatalog-table --hcatalog-
storage-stanza "stored as orcfile"
```

This command will create a new table called TEST10_ORC in the default database with data stored in ORC file format. In most cases, you store the Hive table data in ORC format to make use of the latest performance optimizations, for example, vectorization. This command is quite handy for creating table definitions and loading data into the ORC format in a single step. Once the data is loaded, you can verify its format as follows:

```
hive> describe extended test10_orc;
OK
a                   int
b                   int
```

```
Detailed Table Information Table(tableName:test10_orc, dbName:default, owner:root,
createTime:1465946427, lastAccessTime:0, retention:0, sd:StorageDescriptor(cols:[Field
Schema(name:a, type:int, comment:null), FieldSchema(name:b, type:int, comment:null)],
location:hdfs://sandbox.hortonworks.com:8020/apps/hive/warehouse/test10_orc, inputFormat:org.
apache.hadoop.hive.ql.io.orc.OrcInputFormat, outputFormat:org.apache.hadoop.hive.ql.io.
orc.OrcOutputFormat, compressed:false, numBuckets:-1, serdeInfo:SerDeInfo(name:null,
serializationLib:org.apache.hadoop.hive.ql.io.orc.OrcSerde, parameters:{serialization.
format=1}), bucketCols:[], sortCols:[], parameters:{}, skewedInfo:SkewedInfo(skewedColNam
es:[], skewedColValues:[], skewedColValueLocationMaps:{}), storedAsSubDirectories:false),
partitionKeys:[], parameters:{transient_lastDdlTime=1465946427}, viewOriginalText:null,
viewExpandedText:null, tableType:MANAGED_TABLE)
Time taken: 0.585 seconds, Fetched: 4 row(s)
```

## Importing Selective Data

```
sqoop import --connect jdbc:mysql://localhost/test --table TEST1 --username root --m 1
--where "a>1"
```

With this command, you can import all data from the TEST1 table, where the value of column a is greater than 1. This option provides a way to import a subset of any table.

## Importing Incremental Data

You can also perform incremental imports using Sqoop. Incremental import is a technique that imports only the newly added rows in a table. It is required to add incremental, check-column, and last-value options to perform the incremental import.

- incremental—Used by Sqoop to determine which rows are new. Legal values for this mode include append and lastmodified.

- check-column —To provide the column that needs to checked the determine the candidate rows.

- last-value—This is the maximum value of the last import run.

```
sqoop import --connect jdbc:mysql://localhost/test --username root --table TEST1 --m 1
--incremental append --check-column id –last-value 1000
```

# Apache Nifi

So far the tools we discussed require writing scripts, command-line management, and do not provide any way to track the data as it is transferred into Hadoop. Apache Nifi provides a very easy-to-use, powerful, secure, and trackable way to process and distribute data. It has a very easy-to-use web UI that provides a seamless experience among design, control, management, and monitoring of the data transfer jobs. These jobs are called data flows and, unlike traditional streaming solutions, they can operate in a bidirectional manner. These data flows consists of various processors that provide the logic in terms of the operation that needs to be performed on the data.

Apache Nifi is distributed in the form of a compressed file and the installation just requires unpacking this file in a directory. For the purposes of this discussion, we assume that you have already installed Apache Nifi in your environment.

We will create a simple data flow to read the Twitter data and write it in a file in HDFS.

1.  Log in to Apache Nifi by browsing the URL `http://<nifihost>:9090/nifi`. Figure 6-7 shows the Apache Nifi UI.

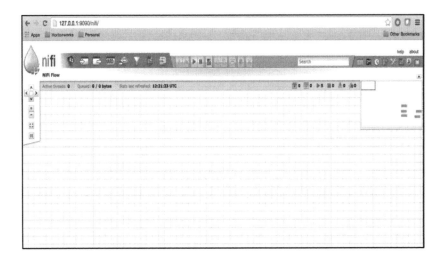

***Figure 6-7.*** *Apache Nifi home page*

2.  Drag the processor icon 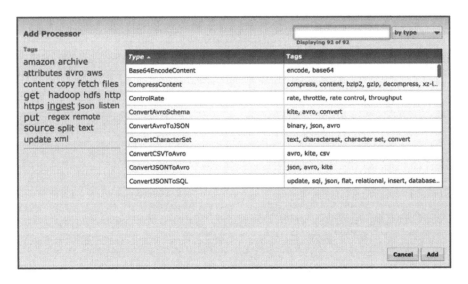 from the toolbar to the grid to open the Add Processor window, as shown in Figure 6-8.

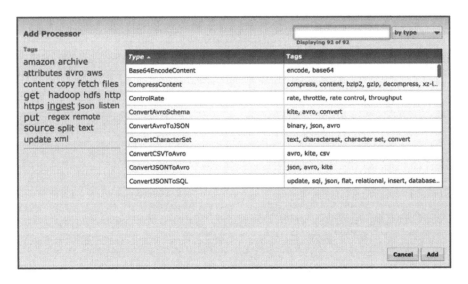

***Figure 6-8.*** *Adding a Nifi processor*

3.   Select the GetTwitter processor (as shown in Figure 6-9) and click on Add. This
     processor is used to read data from the Twitter garden hose. Before we can read
     the data, we need to add some properties to it.

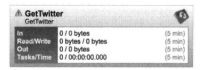

***Figure 6-9.***  *GetTwitter processor*

4.   Right-click on this processor and click Configure.

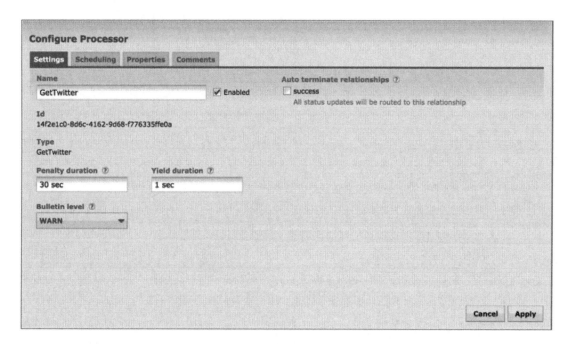

***Figure 6-10.***  *GetTwitter processor configuration window*

5.   Click on the Properties tab and specify the Consumer Key, Consumer Secret,
     Access Token, Access Token Secret, and Terms to Filter on. For example, Hadoop
     (as shown in Figure 6-11).

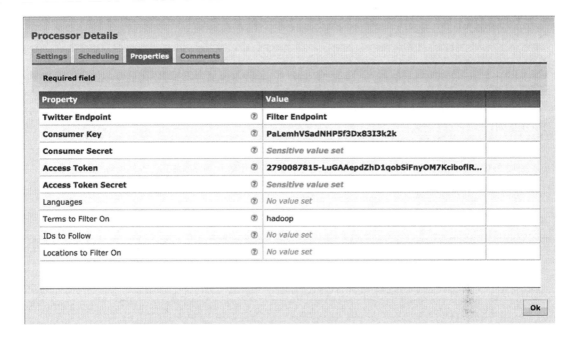

**Figure 6-11.** *GetTwitter processor configuration properties*

6. Now, add another PutHDFS processor and open its configuration properties (see Figure 6-12). You will need to specify the location of the hdfs-site.xml and core-site.xml files and the HDFS directory in which you want to store the tweets.

**Configure Processor**

| Settings | Scheduling | **Properties** | Comments |
| --- | --- | --- | --- |

Required field                                                              ➕ New property

| Property | Value | |
| --- | --- | --- |
| Hadoop Configuration Resources | ⑦ /etc/hadoop/conf/hdfs-site.xml, /etc/hadoop/conf/core-si... | |
| Kerberos Principal | ⑦ *No value set* | |
| Kerberos Keytab | ⑦ *No value set* | |
| **Directory** | ⑦ **/tweets/raw** | |
| **Conflict Resolution Strategy** | ⑦ **fail** | |
| Block Size | ⑦ *No value set* | |
| IO Buffer Size | ⑦ *No value set* | |
| Replication | ⑦ *No value set* | |
| Permissions umask | ⑦ *No value set* | |
| Remote Owner | ⑦ *No value set* | |
| Remote Group | ⑦ *No value set* | |

Cancel    Apply

**Figure 6-12.** *PutHDFS properties*

7.  Once you have added the two processors, the canvas should look like Figure 6-13.

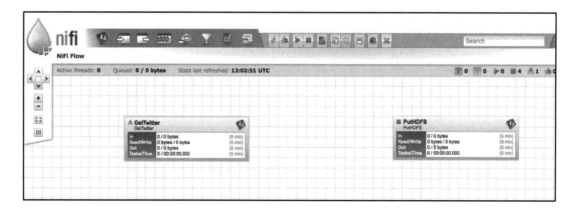

***Figure 6-13.** Apache Nifi processors with no relationship*

8.  We now need to add a relation between these two processors. Click in the middle of the GetTwitter processor and drag toward PutHDFS. You will notice a dotted green line appears between the two processors and the Create Connection window opens (see Figure 6-14).

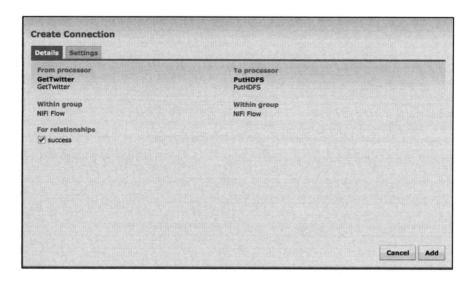

***Figure 6-14.** Create Connection window*

9.  Click on Add to add this connection.

10. As shown in Figure 6-15, we now have a simple data flow ready, which can read the tweets from Twitter and write them to HDFS. Click on the green Play button ▶ in toolbar to start the data flow and save the tweets to HDFS.

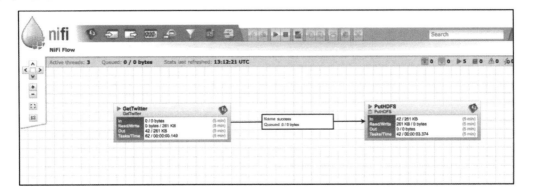

**Figure 6-15.** *A simple data flow example*

> 11.   We can verify the data in HDFS as follows:

```
[root@sandbox ~]# hadoop fs -ls /tweets/raw | wc -l
20574
[root@sandbox ~]# hadoop fs -ls /tweets/raw | head -10
Found 20573 items
-rw-r--r--   1 root hdfs      13929 2016-05-18 09:47 /tweets/raw/10005654822649.json
-rw-r--r--   1 root hdfs       2287 2016-05-18 09:47 /tweets/raw/10006656905343.json
-rw-r--r--   1 root hdfs       2528 2016-02-08 11:05 /tweets/raw/10011382997542.json
-rw-r--r--   1 root hdfs       6469 2016-01-31 08:33 /tweets/raw/10018657101686.json
-rw-r--r--   1 root hdfs       5254 2016-01-31 08:33 /tweets/raw/10021683146427.json
-rw-r--r--   1 root hdfs       9242 2016-05-18 09:48 /tweets/raw/10024390262882.json
-rw-r--r--   1 root hdfs       2580 2016-01-31 08:33 /tweets/raw/10026695152597.json
-rw-r--r--   1 root hdfs       6254 2016-01-31 08:33 /tweets/raw/10029702254017.json
-rw-r--r--   1 root hdfs       7410 2016-01-31 08:33 /tweets/raw/10029707961511.json
[root@sandbox ~]#
```

# Accessing the Data in Hive

By now, you should be familiar with various tools that are available to load data into Hadoop. Most of these tools store the data in the form of a file in HDFS. Landing data in HDFS does not make it accessible in Hive automatically. Hive relies on a table definition, which is stored in Hive Metastore, to access the underlying data from HDFS. Lets look at how we can make the data stored in HDFS available in Hive.

## External Tables

An external table has its metadata stored in Hive Metastore but it does not have full control over the underlying data. The data belonging to external table is stored in HDFS but it can be located in any directory. When you delete an external table, the underlying data is not deleted from HDFS.

These tables are quite useful when you are regularly ingesting files of a similar type in a directory on HDFS. As long as the underlying data has the same format, when you query the external table, it will fetch the latest data from the files on HDFS. In most of the examples, where we have copied the data to HDFS, this data can be made available in Hive by creating an external table on top of these files.

We will now create a table called TEST3 using the following command on one of the text files that we loaded into HDFS in the previous examples.

```
drop table test3;
create external table test3(id INT, age INT)
row format delimited
fields terminated by ','
lines terminated by '\n'
stored as textfile
location '/user/root/TEST3';
0: jdbc:hive2://localhost:10000/default> create external table test3(id INT, age INT)
0: jdbc:hive2://localhost:10000/default> row format delimited
0: jdbc:hive2://localhost:10000/default> fields terminated by ','
0: jdbc:hive2://localhost:10000/default> lines terminated by '\n'
0: jdbc:hive2://localhost:10000/default> stored as textfile
0: jdbc:hive2://localhost:10000/default> location '/user/root/TEST3';
No rows affected (2.029 seconds)
0: jdbc:hive2://localhost:10000/default> select count(*) from TEST3;
INFO  : Tez session hasn't been created yet. Opening session
INFO  :

INFO  : Status: Running (Executing on YARN cluster with App id
application_1465942169140_0016)

INFO  : Map 1: -/-        Reducer 2: 0/1
INFO  : Map 1: 0/1        Reducer 2: 0/1
INFO  : Map 1: 0(+1)/1    Reducer 2: 0/1
INFO  : Map 1: 1/1        Reducer 2: 0/1
INFO  : Map 1: 1/1        Reducer 2: 0(+1)/1
INFO  : Map 1: 1/1        Reducer 2: 1/1
+--------+--+
|  _c0   |
+--------+--+
| 32770  |
+--------+--+
1 row selected (13.368 seconds)
0: jdbc:hive2://localhost:10000/default>
```

## Load Data Statement

You can use the LOAD DATA statement if you want to copy the data into an existing table definition in Hive. The LOAD DATA statement is simply a copy/move operation at the file level. Here is the syntax of the LOAD DATA command:

```
LOAD DATA INPATH 'filepath' [OVERWRITE] INTO TABLE tablename [PARTITION clause];
```

When you execute a LOAD DATA command, the file stored in filepath is copied to the directory specified in the table definition of the target table. We will now revisit the example from "Create External Table" section, to first create a TEST4 table and then load the file using the LOAD DATA command.

```
0: jdbc:hive2://localhost:10000/default> CREATE TABLE TEST4(id INT, age INT) STORED AS
TEXTFILE LOCATION '/tmp/root/TEST4';
No rows affected (1.974 seconds)
0: jdbc:hive2://localhost:10000/default> LOAD DATA INPATH '/user/root/TEST3/test.csv' into
TABLE TEST4;
INFO  : Loading data to table default.test4 from hdfs://sandbox.hortonworks.com:8020/user/
root/TEST3/test.csv
INFO  : Table default.test4 stats: [numFiles=0, numRows=0, totalSize=0, rawDataSize=0]
No rows affected (2.412 seconds)
0: jdbc:hive2://localhost:10000/default> SELECT COUNT(*) FROM TEST4;
INFO  : Session is already open
INFO  : Tez session was closed. Reopening...
INFO  : Session re-established.
INFO  :

INFO  : Status: Running (Executing on YARN cluster with App id
application_1465942169140_0017)

INFO  : Map 1: -/-        Reducer 2: 0/1
INFO  : Map 1: 0/1        Reducer 2: 0/1
INFO  : Map 1: 0(+1)/1    Reducer 2: 0/1
INFO  : Map 1: 1/1        Reducer 2: 0/1
INFO  : Map 1: 1/1        Reducer 2: 0(+1)/1
INFO  : Map 1: 1/1        Reducer 2: 1/1
+--------+--+
|  _c0   |
+--------+--+
| 32770  |
+--------+--+
1 row selected (12.613 seconds)
0: jdbc:hive2://localhost:10000/default>
```

# Loading Incremental Changes in Hive

Loading data into Hadoop is a continuous task. Once you have loaded a large amount of data from a source system initially, you can bring the changes in the form of regular batch runs. In case of Hive, this is done by bringing in new data in the form of delta files and adding new partitions to the table. However, you cannot modify the data in existing partitions. As a part of Stinger.Next initiative, the community is adding ACID functionality to Hive. With this core functionality of insert/update, we also have a set of streaming APIs that allow a continuous ingestion of the data to tables in Hive.

# Hive Streaming

The Hive streaming API is mainly used with Storm as a Hive Bolt. It breaks down a stream of data into smaller batches. The incoming data can be continuously committed in small batches of records into an existing Hive partition or table. Once data is committed it becomes immediately visible to all Hive queries initiated subsequently. As mentioned earlier, this streaming functionality is based on insert/update support.

There are currently some limitations on the Hive streaming API:

- The target table must be bucketed

- The streaming API only provides support for streaming delimited input data (such as CSV, tab separated, etc.) and JSON (strict syntax) formatted data

- The target table must be stored in ORC format

- You must set the required parameters to enable ACID functionality

  - `hive.txn.manager = org.apache.hadoop.hive.ql.lockmgr.DbTxnManager`

  - `hive.compactor.initiator.on = true`

  - `hive.compactor.worker.threads > 0`

The actual implementation of Hive streaming requires a Storm Bolt to be written in Java, which is beyond the scope of this book.

# Summary

In this chapter, we looked at various options to load data into Hive. In most cases, loading data in Hive is a two-stage process. All data is first ingested to HDFS, then its metadata is added to Hive Metastore. There are many options when it comes to using a tool to ingest the data in HDFS. These tools have been built for various use cases. Apache Nifi is commonly used to ingest almost all types of data these days. Its out-of-the-box unique features (such as provenance, security, and ease of management) make it a very suitable tool for enterprise data ingestion into the Hadoop data lake. As more and more companies use Hadoop for real-time processing use cases, such use cases require continuous data ingestion from operational systems. Hive streaming, although still not fully ready for production, provides this functionality through Hive ACID. Some of the RDBMS vendors have also created plug-ins for their Change Data Capture (CDC) technologies like Oracle GoldenGate, Attunity, etc., to load continuous changes to Hive tables. However, there is still a lot of work that needs to be done in this space to make real-time changes accessible and effective.

# CHAPTER 7

# Querying Semi-Structured Data

Hive would not be much of a useful data warehouse tool without the ability to query data. Luckily, querying and providing schema-on-read capabilities at scale is the core foundation for Hive use cases. The power Hive provides is the ability to translate a large variety of data formats as well as the ability to customize translations to fit your unique business needs. Hive adapts to your data formats instead of the other way around. This is the core foundation for a data-driven organization.

Hive accomplishes this through HCatalog, as described earlier, but also through unique storage and load capabilities. You will find many parts of Hive familiar if you are already well-versed in existing query languages, but you will also find nuances which extend query capabilities and schemas well beyond what is available in a traditional RDBMS.

The Hadoop noise machine was fond of referring to data as structured, semi-structured, or non-structured. Structured data always referred to data represented in rows and columns. This is what was most familiar to data analysts, especially professionals working with traditional transactional systems like point-of-sales or inventory management. Semi-structured data refers to a gray line between columns and rows and maybe something more exotic like key-value pairs, arrays, or nested data. Maybe the number of columns in the data structure was dynamic, or maybe there were multiple values in a single column. This data felt like traditional data but its representation was much different. Examples of this data include XML, HL7, and JSON. Here is an actual tweet represented as a JSON file (the file is too long to show in its entirety, so this is an abbreviated version):

```
{
  "created_at": "Wed Sep 23 01:19:54 +0000 2015",
  "id": 646494164109029400,
  "id_str": "646494164109029376",
  "text": "@StarksAndSparks \"I'm not!\" He laughs and shrugs. \"I'm all bone.\"",
  "source": "<a href=\"http://twitter.com/download/iphone\" rel=\"nofollow\">Twitter for iPhone</a>",
  "truncated": false,
  "in_reply_to_status_id": 646222681067622400,
  "in_reply_to_status_id_str": "646222681067622400",
  "in_reply_to_user_id": 3225146093,
  "in_reply_to_user_id_str": "3225146093",
  "in_reply_to_screen_name": "StarksAndSparks",
  "user": {
    "id": 3526755197,
    "id_str": "3526755197",
    "name": "smoll steve",
    "screen_name": "hellatinysteve",
    "location": "",
```

```
  "url": null,
  "description": "like a chihuahua who thinks he's a pitbull. did someone say napoleon
  complex?",
  "protected": false,
  "verified": false,
  "followers_count": 117,
  "friends_count": 56,
  "listed_count": 3,
  "favourites_count": 155,
  "statuses_count": 1831,
  "created_at": "Wed Sep 02 20:26:36 +0000 2015",
  "utc_offset": null,
  "time_zone": null,
  "geo_enabled": true,
  "lang": "en",
  "contributors_enabled": false,
  "is_translator": false,
  "profile_background_color": "CODEED",
  "profile_background_image_url": "http://abs.twimg.com/images/themes/theme1/bg.png",
  "profile_background_image_url_https": "https://abs.twimg.com/images/themes/theme1/bg.png",
  "profile_background_tile": false,
  "profile_link_color": "0084B4",
  "profile_sidebar_border_color": "CODEED",
  "profile_sidebar_fill_color": "DDEEF6",
  "profile_text_color": "333333",
  "profile_use_background_image": true,
  "profile_image_url": "http://pbs.twimg.com/profile_images/639178684478394368/0f3yigOF_
  normal.jpg",
  "profile_image_url_https": "https://pbs.twimg.com/profile_images/639178684478394368/
  0f3yigOF_normal.jpg",
  "profile_banner_url": "https://pbs.twimg.com/profile_banners/3526755197/1441227570",
  "default_profile": true,
  "default_profile_image": false,
  "following": null,
  "follow_request_sent": null,
  "notifications": null
},
"geo": null,
"coordinates": null,
"place": null,
"contributors": null,
"retweet_count": 0,
"favorite_count": 0,
"entities": {
  "hashtags": [],
  "trends": [],
  "urls": [],
  "user_mentions": [
    {
      "screen_name": "StarksAndSparks",
      "name": "Tony Stark.",
```

```
      "id": 3225146093,
      "id_str": "3225146093",
      "indices": [
         0,
         16
      ]
   }
],
"symbols": []
},
"favorited": false,
"retweeted": false,
"possibly_sensitive": false,
"filter_level": "low",
"lang": "en",
...
```

As you can see, there is a wealth of information in every tweet. The power of Hadoop is the ability to store a raw file like a JSON tweet like you would store a file on any filesystem, but then be able to use Hive to create a schema over the directory that allows you to query attributes of the raw data. You can store all the data but only query the data you need.

Semi-structured data could also be associated with syslog or application event log files. Finally, there was unstructured data in the form of images, OCR, PDF, or spatial data. Unstructured data was complex data where potentially the structure was not in columns, rows, or arrays, but was in the byte patterns in an image of a cat on the Internet or a rib cage in a X-ray. The truth of the matter is no data is patternless. What matters is the algorithm used to detect the pattern. Granted, the pattern may change during moments of ingest or may not be readily or easily detectable, but all data still has a pattern and it is up to developers to glean those patterns using all the tools at their disposal, and it is up to the tools analyzing the data to have the flexibility to accommodate the potential range of patterns.

This chapter primarily focuses on the semi-structured data and how we can leverage this data in Hive for reporting and analytics. We look at practical data like clickstream, JSON, and server log data. By the end of the chapter, you should have a good handle on how to ingest and create schemas on this data as well as understand the ingest and translation tools available for expanding data you can use in Hive.

# Clickstream Data

A common use case is leveraging clickstream data to analyze and predict customer behavior. Some questions you can answer through the data include:

- Which page is most popular?

- Which page do most users drop off from?

- Are users staying on a particular page longer than others?

- What is the most common navigation path?

As a business you can use the answers to these questions to help promote certain items or customize your web page to fit behavior patterns. Furthermore, if you are able to capture this data in real-time, you have the ability to get immediate feedback and make corrections when necessary. Marketing and content creators can receive instant feedback on changes and promotions and react to them in near real-time. Storing this data in HDFS and querying through Hive can also provide for trending analysis for forecasting and predictive analytics.

There is no shortage of clickstream tools available. Many of these tools, such as Google Analytics, are cloud-based. Using such tools, you are able to gather the data and view results in canned graphs and charts. What a Hive plus HDFS option provides for you is the ability to own your own data and potentially enrich the data with other internal data such as internal product or sales data. As we will see, the effort to ingest, store, and then visualize the data is relatively easy, and it is a project that many companies start with when beginning their Hadoop journey.

Clickstream data is normally stored as log files usually in a directory on a web server. The most common way to ingest these files is through an application such as Apache Flume or Apache Nifi. Setting up and configuring Apache Flume is out of scope for this chapter so we will primarily focus on manually copying the log to HDFS. In our example we will ingest raw Wikipedia clickstream data. You can download the data from `https://figshare.com/articles/Wikipedia_Clickstream/1305770`. There are four data sets with a total data set size of 2.37 GB. It does not matter which data set you choose and choosing the entire data set is also fine.

---

■ **Note**　Apache Flume is an easy-to-use method to ingest running log files into HDFS. Flume runs as an agent and in Flume you create sources for log processing. You can have multiple sinks, which perform the processing as well as multiple agents with guaranteed delivery. For more information, visit the Apache Flume site at `https://flume.apache.org/`.

---

The Wikipedia data consists of web site traffic during the month of January 2015. The data focuses on page referrals, that is the current page the user was on and where the user went. This referral can be using a search engine or clicking on a link on a page. Let's take a look a sample from one of the data sets:

| | | | | |
|---|---|---|---|---|
| 1758827 | 2516600 | 154 | !Kung_people | !Kung_language |
| 22980 | 2516600 | 74 | Phoneme | !Kung_language |
| | 2516600 | 20 | other | !Kung_language |
| 261237 | 2516600 | 21 | The_Gods_Must_Be_Crazy | !Kung_language |
| 247700 | 2516600 | 12 | Xu_language | !Kung_language |
| | 2516600 | 29 | other-wikipedia | !Kung_language |
| 1383618 | 2516600 | 33 | Mama_and_papa | !Kung_language |
| 7863678 | 2516600 | 12 | List_of_endangered_languages_in_Africa | !Kung_language |
| 524854 | 2516600 | 20 | Alveolar_clicks | !Kung_language |
| 34314219 | 2516600 | 11 | Ekoka_!Kung | !Kung_language |
| 27164415 | 2516600 | 100 | Contents_of_the_Voyager_Golden_Record | !Kung_language |
| 524853 | 2516600 | 21 | Palatal_nasal | !Kung_language |
| 17333 | 2516600 | 45 | Khoisan_languages | !Kung_language |
| 713020 | 2516600 | 56 | Jul'hoan_dialect | !Kung_language |
| | 29988427 | 300 | other-empty | !Women_Art_Revolution |
| | 29988427 | 93 | other-google | !Women_Art_Revolution |
| | 29988427 | 24 | other-wikipedia | !Women_Art_Revolution |
| 420777 | 29988427 | 14 | Zeitgeist_Films | !Women_Art_Revolution |
| 6814223 | 29988427 | 23 | Lynn_Hershman_Leeson | !Women_Art_Revolution |
| 1686995 | 29988427 | 27 | Carrie_Brownstein | !Women_Art_Revolution |
| | 64486 | 650 | other-empty | !_(disambiguation) |
| | 64486 | 226 | other-google | !_(disambiguation) |
| | 64486 | 23 | other-wikipedia | !_(disambiguation) |
| 600744 | 64486 | 14 | !!! | !_(disambiguation) |
| 7712754 | 64486 | 237 | Exclamation_mark | !_(disambiguation) |

To give you an idea of the size, the full data set contains 22 million referrer article pairs, but is still only a sampling of the 4 billion total requests made in January! The data set has six fields:

- prev_id—If the referer does not correspond to an article in the main namespace of English Wikipedia, this value will be empty. Otherwise, it contains the unique MediaWiki page ID of the article corresponding to the referrer, i.e., the previous article the client was on.

- curr_id—The unique MediaWiki page ID of the article the client requested.

- n—The number of occurrences of the *(referer, resource)* pair.

- prev_title—The result of mapping the referer URL to the fixed set of values described above.

- curr_title—The title of the article the client requested.

- type

  - "link" if the referer and request are both articles and the referer links to the request.

  - "redlink" if the referer is an article and links to the request, but the request is not in the production enwiki.page table.

  - "other" if the referer and request are both articles but the referer does not link to the request. This can happen when clients search or spoof their referer.

If you notice, not all fields are present in every row of the data and this can be a problem when ingesting data through traditional ETL processing. Data that's NULL in nature still has to be accounted for and your table will need to be defined for all possible fields whether they exist or not. When using HDFS and Hive, we will ingest the data first. Once the data is ingested, we will create the schema. This is the value of schema-on-read and it is part of what makes a Hive data warehouse development much more agile than traditional data warehousing development.

## Ingesting Data

The first step is data ingestion and as mentioned before, we will manually simulate what would normally be a log-streaming ingestion process. You should have downloaded a compressed file with a name similar to 2015_01_clickstream.tsv.gz. If you only download one data set, the compressed file is about 330 MB. If you were to uncompress it, the file would explode to over 1 GB. Files like clickstream data compress well and you can normally expect over 70% compression. What is useful is there is no need to uncompress these files when storing them in HDFS.

---

■ **Caution**    Accessing files natively in Hadoop with compression works for GZIP extensions but not for ZIP extensions. If you try to query data stored in files with a .ZIP extension, you will only get null values. If you have to work with .ZIP files, there are some options to wrap a ZIP file reader around MapReduce InputFormat.

---

To begin ingesting data, first go to Ambari and create a landing directory in HDFS. This is where we will upload the file prior to creating a table in Hive. We can do this through the Ambari HDFS view. Figure 7-1 shows you how to get to the HDFS view.

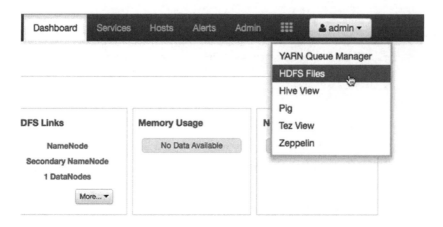

***Figure 7-1.*** *HDFS Files view*

Once you're in the HDFS Files view, navigate to the /tmp directory. You can create the Wiki clickstream directory anywhere you would like, but for the purposes of this exercise, we will use /tmp. Once in /tmp, click on Create Directory and create a directory called wikiclickstream. Figure 7-2 shows the New Directory option.

| / tmp | | | | | | | + New directory | ± Upload |
|---|---|---|---|---|---|---|---|---|

| ■ tmp ☑ | | | | | | | Search File Names | 🔍 |
|---|---|---|---|---|---|---|---|---|

| Name | ∨ | Size | > | Last Modified | > | Owner | > | Group | > | Permission | > | Asc | Name ▾ | | ☐ ▾ |
|---|---|---|---|---|---|---|---|---|---|---|---|---|---|---|---|
| ■ .. | | | | | | | | | | | | | | | |
| ■ entity-file-history | | - | | 2016-03-14 09:19 | | hdfs | | hdfs | | -rwxr-xr-x | | | | | ☐ |
| ■ hive | | - | | 2016-04-20 09:14 | | ambari-qa | | hdfs | | -rwx-wx-wx | | | | | ☐ |
| ■ tweets_staging | | - | | 2016-04-20 14:49 | | nifi | | hdfs | | -rwxr-xr-x | | | | | ☐ |

***Figure 7-2.*** *Creating a new directory*

Follow the prompts and you should now see a directory called wikiclickstream in the /tmp directory. Click on the wikiclickstream directory to navigate into it. We will now upload the compressed clickstream data by clicking on the Upload button and browsing to the file we previously downloaded. Figure 7-3 shows the Upload button and the downloaded file. Notice that the file still has the compressed .GZ extension.

| / tmp / wikiclickstream | | | | | | | + New directory | ± Upload |
|---|---|---|---|---|---|---|---|---|

| ■ wikiclickstream ☑ | | | | | | | Search File Names | 🔍 |
|---|---|---|---|---|---|---|---|---|

| Name | ∨ | Size | > | Last Modified | > | Owner | > | Group | > | Permission | > | Asc | Name ▾ | | ☐ ▾ |
|---|---|---|---|---|---|---|---|---|---|---|---|---|---|---|---|
| ■ .. | | | | | | | | | | | | | | | |
| ■ 2015_01_clickstream.tsv.gz | | 313.3 MB | | 2016-04-30 13:28 | | admin | | hdfs | | -rw-r--r-- | | | | | ☐ |

***Figure 7-3.*** *Uploading a clickstream file*

The data is now loaded into HDFS. Our data set is small but this could potentially be a multi-terabyte file loaded through an automated batch process or a real-time streaming process. Click on the file to view a sample of the contents. Notice that HDFS automatically uncompresses the file for viewing. Figure 7-4 shows the file's contents.

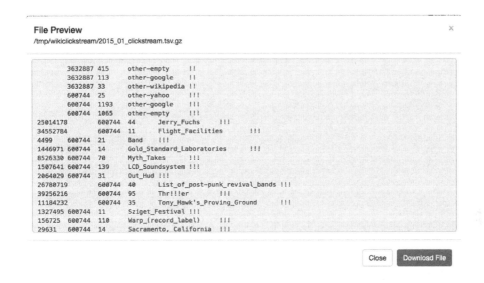

**Figure 7-4.** *Contents of clickstream file*

The only thing left to do is create metadata on the file. Essentially we will build a view or virtual table that points to the file so that you can run Hive queries against the data. To do this, we can create our table DDL in a script and run it in HiveCL, run our DDL directly in HiveCL, or execute the script in the Hive view. For our data, we will use the Hive view. Navigate to the Hive view, which is in the same location as the HDFS view. In the query editor, execute the following command:

```
CREATE DATABASE clickstream;
```

Figure 7-5 shows the command as well as the newly created database. You will need to refresh the database explorer to see the clickstream database.

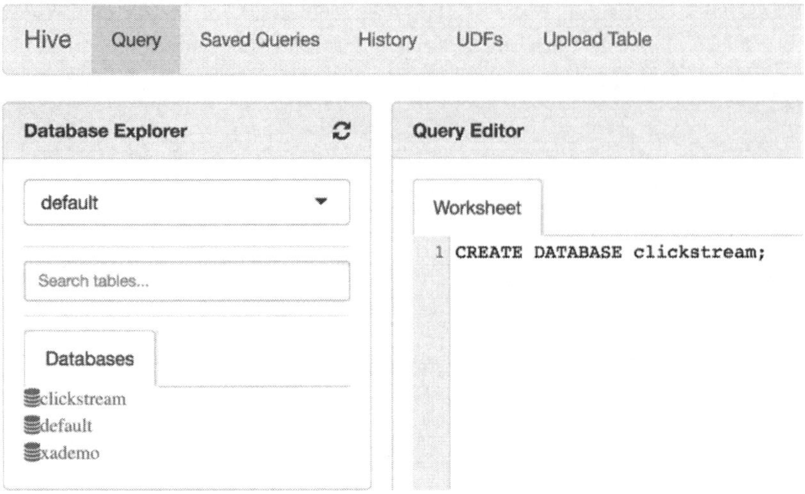

**Figure 7-5.** *Creating a clickstream database*

Once you see it, move your database from `default` to `clickstream`. You can do this by selecting the database from the drop-down menu or executing the following code in the query editor:

```
USE clickstream;
```

Creating a new database specifically for the clickstream data will help us organize our project. Notice that creating a database in Hive is simple and straightforward. You do not need to allocate any memory or storage requirements and there are no files or settings associated with the database. This is because the database is only a metadata container for any tables you create under it.

## Creating a Schema

Now that we have created the database, let's create the table. Copy and execute the following script to create the `wikilogs` table.

```
CREATE TABLE wikilogs (
        previous_id     STRING,
        current_id      STRING,
        no_occurences   INT,
        previous_title  STRING,
        current_title   STRING,
        type            STRING)
ROW FORMAT DELIMITED
FIELDS TERMINATED BY '09'
STORED AS textfile;
```

The Hive `CREATE TABLE` statement should look familiar to anyone who knows SQL. The primary difference with this `CREATE TABLE` command are the last three commands. The `ROW FORMAT DELIMITED` command lets Hive know that there is a delimiting character in the file and each field is separated by a Tab (09 is the ASCII value for the Tab command). Fields can be separated by almost any character and this would be represented in the `CREATE TABLE` statement. The `STORED AS` command tells Hive how to store the file. In this case we will store it as a basic text file. In the real world, you would store the data in a more performant file format such as an ORC file. These file formats are discussed in the chapter on performance tuning.

# Loading Data

The next step is to load the data into the Hive table. You really are not as much loading data as you are moving the file to a Hive directory. In this example, we created a Hive managed table, which means Hive will also manage the data and the data will be removed if the table is dropped. We could have created a Hive external table and pointed the table to a location in HDFS. With an external table, the data is not removed when the table is dropped. Loading the data as simple as executing a LOAD statement.

```
LOAD DATA INPATH '/tmp/wikiclickstream/2015_01_clickstream.tsv.gz' OVERWRITE INTO TABLE
wikilogs;
```

You will want to change the filename and directory path to the one that is correct for your system. It is key to understand that the LOAD command does not perform any transformations on the data, but instead; the LOAD command simply copies data to the location specified or defaulted in the table DDL. The OVERWRITE command simply deletes any existing data associated with the table and uses the new file data in the LOAD command. If an old file exists with the same name, the new file will replace the old one.

# Querying the Data

After executing the LOAD command, you now have data available in your wikilogs table. Let's first do some cleanup by eliminating some unnecessary columns. For our purposes, we do not need the first two columns since these are page identifications without values. We are primarily concerned with only the page the user was on, the page he went to, and how many times that sequence occurred for all users. We could have defined our table originally without the columns but maybe another group in our company had a need for that data. For our group we will create a simple view to limit those two columns. Execute the following HiveQL in the query editor window.

```
CREATE VIEW wikilogs_view (no_occurences, previous_title, current_title)
AS SELECT no_occurences, previous_title, current_title FROM wikilogs;
```

Now that we have a view, we can begin to ask some questions about the data. Let's first find the link with the highest number of occurrences. Execute the following query, but keep in mind that depending on your data set size, it could take a while to return the results. Up until now we have done no performance tuning and since we are on the sandbox we are not taking advantage of any distributed, parallel processing.

```
SELECT * FROM wikilogs_view
SORT BY no_occurences DESC;
```

Figure 7-6 shows the results.

**Figure 7-6.** *Results of sorting by no_occurences*

The data tells us that by far the most common link occurrence is the Wiki home page. This makes sense considering the nature of the data. The previous_title field includes common search sources and can help us determine which search tools people use most to find data on Wikipedia. The values are defined as follows:

- An article in the main namespace of English Wikipedia ➤ The article title
- Any Wikipedia page that is not in the main namespace of English Wikipedia ➤ other-wikipedia
- An empty referer ➤ other-empty
- A page from any other Wikimedia project ➤ other-internal
- Google ➤ other-google
- Yahoo ➤ other-yahoo
- Bing ➤ other-bing
- Facebook ➤ other-facebook
- Twitter ➤ other-twitter
- Anything else ➤ *Other-Other*

Based on these values, we can answer questions such as, "What is the most frequently linked Wiki page on Facebook"? Let's find out by executing the following SQL.

```
SELECT * FROM wikilogs_view
WHERE previous_title = 'other-facebook'
SORT BY no_occurences DESC;
```

Figure 7-7 shows the results.

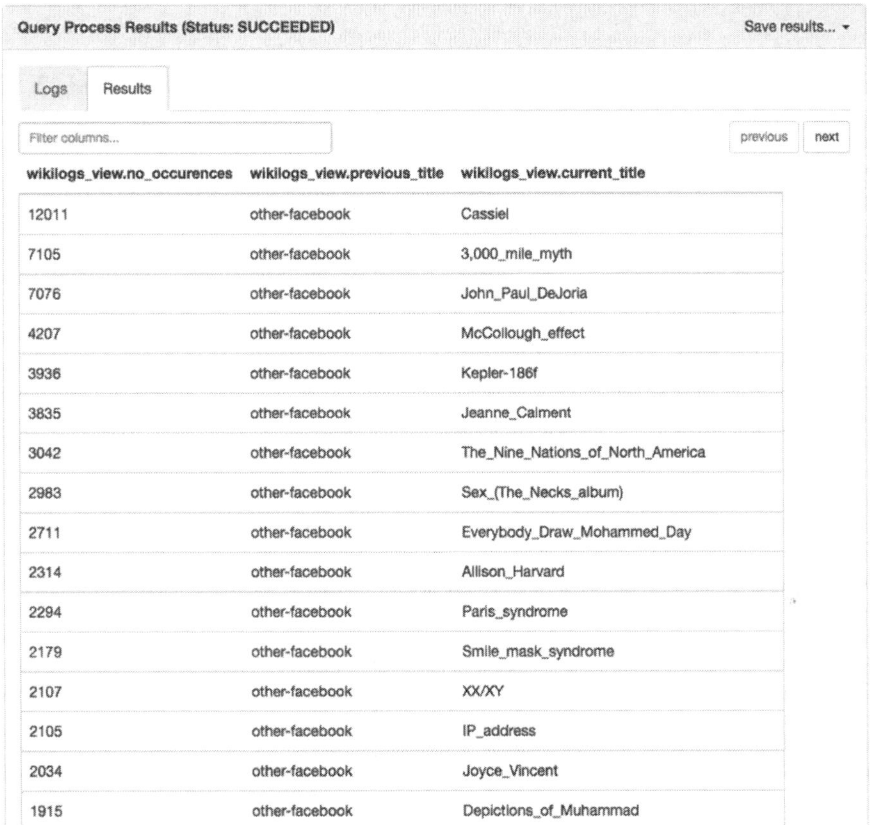

| wikilogs_view.no_occurences | wikilogs_view.previous_title | wikilogs_view.current_title |
| --- | --- | --- |
| 12011 | other-facebook | Cassiel |
| 7105 | other-facebook | 3,000_mile_myth |
| 7076 | other-facebook | John_Paul_DeJoria |
| 4207 | other-facebook | McCollough_effect |
| 3936 | other-facebook | Kepler-186f |
| 3835 | other-facebook | Jeanne_Calment |
| 3042 | other-facebook | The_Nine_Nations_of_North_America |
| 2983 | other-facebook | Sex_(The_Necks_album) |
| 2711 | other-facebook | Everybody_Draw_Mohammed_Day |
| 2314 | other-facebook | Allison_Harvard |
| 2294 | other-facebook | Paris_syndrome |
| 2179 | other-facebook | Smile_mask_syndrome |
| 2107 | other-facebook | XX/XY |
| 2105 | other-facebook | IP_address |
| 2034 | other-facebook | Joyce_Vincent |
| 1915 | other-facebook | Depictions_of_Muhammad |

***Figure 7-7.*** *Top links from Facebook*

The top three links are Cassiel, 3,000_mile_myth, and John_Paul_DeJoria. It is difficult based on the data we have to speculate why these particular links from Facebook were popular in January of 2015, but it would be interesting to dive into other data during the period to determine why there is a significant gap between the number one spot and the other two positions. Additional data that may help us find details about the individual Facebook postings, data from the other sources such as Google or Twitter, and geographical data.

Of course your business's clickstream data would contain considerably more fields than the Wikipedia example shown here, but the ingestion, storage, and querying process remains the same. Many companies will ingest clickstream data and merge it with internal marketing data for custom ad placements or strategic promotional offers. Another use case is to include syslog data from the application servers so that operations teams can better identify web page errors resulting from application or hardware failures. The ability to store more data at faster rate opens doors into more efficient and proactive maintenance as well as driving quicker and more cost-effective business decisions.

# Ingesting JSON Data

In this next section, we take a quick look at a more complex data type. We ingest JSON data, which is popularly known as the format of Twitter data. JSON (JavaScript Object Notation) is one of the most widely used formats for transmitting application data. It is a popular open standard similar to XML. Like XML, it is based on an attribute\value pairing. The value can be almost anything including a single element, long text, or even maps and arrays. JSON can also be heavily nested or have dynamic attributes, and this can cause problems with standard ETL processes. The popularity of JSON has produced a number of applications and programming languages for reading and parsing JSON data. Some dataflow products will even convert any incoming data to JSON prior to ingesting into HDFS or a NoSQL database.

---

■ **Note**    For an advanced (and more fun) Twitter feed example, I suggest the tutorial that uses Apache Nifi for connecting to the Twitter garden hose. Tweets are sent into Nifi and routed to a Solr Banana dashboard and then also routed to HDFS for longer-term storage. The whole thing can be set up in less than an hour. You will find all the instructions you need on the Hortonworks Community Connection at `https://community.hortonworks.com/content/kbentry/1282/sample-hdfnifi-flow-to-push-tweets-into-solrbanana.html`.

---

The example we use in this chapter consists of ingesting random JSON files and then building a table in Hive so that we can query the data. The ingestion phase is straightforward but there will be some things we need to consider when building the tables and querying the data. The decisions we make will impact query performance. We discuss all the possible options as we go through the example.

Before we do anything, we need data. Luckily, getting JSON data is simple. The method we used for this example is the JSON generator at `http://beta.json-generator.com/`. The site randomly creates data for any JSON template you upload. For simplicity's sake, we are going to use the default template. When you generate data it will create a listing of multiple JSON elements. Each element or block starts with _id. We have gone ahead and separated these into different JSON files named json1, json2, and json3. Here is the content of json1:

```
{
  "_id": "5774245438f862f0b8121f41",
  "index": 5,
  "guid": "580ff472-9036-40b2-aa3c-9085f305d6b4",
  "isActive": false,
  "balance": "$2,252.98",
  "picture": "http://placehold.it/32x32",
  "age": 36,
  "eyeColor": "brown",
  "name": {
```

```
    "first": "Lori",
    "last": "Pacheco"
  },
  "company": "LUDAK",
  "email": "lori.pacheco@ludak.net",
  "phone": "+1 (891) 415-2253",
  "address": "290 Rochester Avenue, Cannondale, Guam, 7856",
  "about": "Qui fugiat nostrud qui laborum Lorem excepteur. Minim exercitation esse mollit
  irure fugiat eiusmod proident sit Lorem incididunt. Dolor ex ipsum tempor est eu duis
  exercitation. Enim ea ullamco mollit proident labore eiusmod excepteur magna Lorem anim.",
  "registered": "Tuesday, February 10, 2015 8:07 AM",
  "latitude": "75.805649",
  "longitude": "138.091539",
  "tags": [
    "ullamco",
    "in",
    "voluptate",
    "reprehenderit",
    "sunt"
  ],
  "range": [
    0,
    1,
    2,
    3,
    4,
    5,
    6,
    7,
    8,
    9
  ],
  "friends": [
    {
      "id": 0,
      "name": "Byrd Meyers"
    },
    {
      "id": 1,
      "name": "Weeks Miles"
    },
    {
      "id": 2,
      "name": "Marquez Pace"
    }
  ],
  "greeting": "Hello, Lori! You have 6 unread messages.",
  "favoriteFruit": "banana"
}
```

Notice that values for things like name, friends, and range have listings or an array of elements. It's this flexible structure that makes JSON so powerful but yet so difficult to ingest into relational systems. Relational systems also struggle with querying this type of data and can make the data difficult to tune for performance. Considering HDFS is a filesystem, data ingestion is trivial, at least for our purposes since we do not require any real-time or automated processes.

---

■ **Note**   When creating JSON data for query consumption, it is better to make sure all unnecessary character data is removed from the file and your JSON schema is properly formed. As mentioned previously, we find it helpful to generate random JSON data with the JSON generator `http://beta.json-generator.com` and then paste the JSON into a JSON editor, which will verify the format as well as flatten out the JSON. A good online editor can be found at `http://www.jsoneditoronline.org/`.

---

Just like we did with the clickstream data, we will create a directory in HDFS to store the JSON data. We chose the same `tmp` directory and created an additional directory called `json_data`. Once created, you will want to open the directory and add the `json1`, `json2`, and `json2` files. Figure 7-8 shows the files in the `json_data` directory.

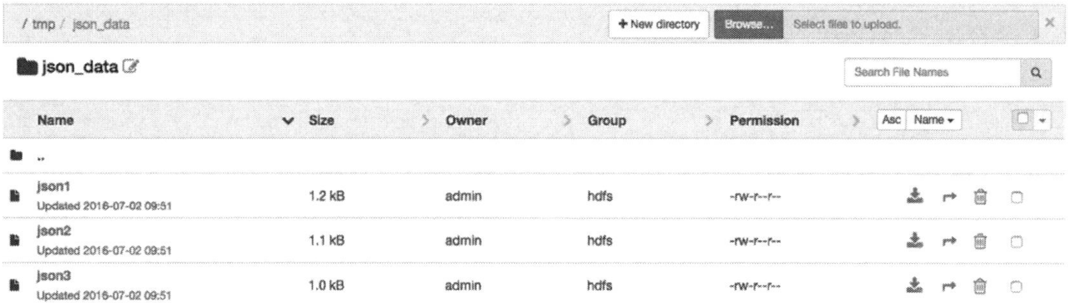

*Figure 7-8.*  *Adding JSON files to HDFS*

## Querying JSON with a UDF

Once the files are added, our next step is to create a schema on the data. This is where it gets interesting and where you will need to make a decision on how you will go about querying the data. There are two primary ways to access JSON data. You can use a built-in UDF (user defined function) or you can use a built-in or publicly available JSON SerDe. The method you decide for JSON access will define how you store the data and the schema you apply to the data.

Let's use the UDF method first. The UDF method is the simplest because it uses native Hive functions and requires a simple schema. The first step is to create a table to store the JSON data. This table will consist of a single string column to represent the entire JSON data. We created a database named `json_data` and will create the table in this database. You can use whichever database you choose. Execute the following command either in the Hive view or from the command line:

```
CREATE TABLE json_table (
json string);
```

As you can see, the table only has one column and we have defined the column as a string. The next step is to load the JSON data into this table and store all the data as a single string column. Execute the following command from the Hive view or from the command line.

```
LOAD DATA INPATH '/tmp/json_data/json1' INTO TABLE json_table;
```

In this example, we created a directory in `tmp` called `json_data` and uploaded the `json1` file. This `LOAD` statement takes that file and loads it into a table called `json_udf`. To query all the data in the `json_udf` table, you execute the following query, which utilizes the `get_json_object` user defined function:

```
select get_json_object(json_table.json, '$') from json_table;
```

This will return all the data from the `json_udf` table. If you want to select multiple values, you need to execute a `select` statement for each value. The following is an example of selecting multiple values:

```
select get_json_object(json_table.json, '$.balance) as balance,
          get_json_object(json_table, '$.gender) as gender,
          get_json_object(json_table.json, '$.phone) as phone,
          get_json_object(json.table.json, '$.friends.name) as friendname
          from json_udf;
```

This query will bring back the balance, gender, phone, and name of the friend from the `json_udf` table. As you can start to notice, this query could begin to get complicated as additional rows are selected and as the data becomes increasingly nested. The table also must be accessed each time a row is needed and this iterative processing can cause significant performance overhead. The benefit of the UDF is that it is built into Hive and you do not have to create a complex schema or try to define a schema based on the content and format of the JSON data. The choice is yours and the `get_json_object` is a good choice for small JSON data sets or when you only need to retrieve a few key attributes.

## Accessing JSON Using a SerDe

By far the most flexible and scalable means to access JSON data is through a SerDe. SerDe is short notation for serializer\deserializer and is a means for Hive to read data from a table and write it out in any customizable format. Developers write SerDes so that Hive can interpret varying file formats.

One such format is JSON. Although there are a few, the most commonly used SerDe for reading JSON data in Hive was written by Roberto Congiu and it can be found on GitHub at `https://github.com/rcongiu/Hive-JSON-Serde`. You will need to follow the instructions to compile the JAR files or you can download the binaries directly. In any case, you will need to place the JAR file in a location accessible from within your Hive environment. In this example, the JAR file is located at `/usr/local/Hive-JSON-Serde/json-serde/target/json-serde-1.3.8-SNAPSHOT-jar-with-dependencies.jar`.

Once you have the JAR file in place, you can start Hive through the command line or through your Ambari view. After Hive starts and before we execute any queries, we have to tell Hive which SerDe we are using by issuing the `ADD` command. Type the following into your Hive line and execute the command:

```
ADD JAR /usr/local/Hive-JSON-Serde/json-serde/target/json-serde-1.3.8-SNAPSHOT-jar-with-
dependencies.jar;
```

You also have the option of adding this command to your `\hiverc` file so that it's available each time Hive starts.

Now that Hive is aware of the SerDe, you can create a table to hold the JSON data. Run the following DDL from the command line (the best method is to refer to a HiveQL file) or from the Ambari view:

```
CREATE TABLE json_serde_table (
  id string,
  about string,
  address string,
  age int,
  balance string,
  company string,
  email string,
  eyecolor string,
  favoritefruit string,
  friends array<struct<id:int, name:string>>,
  gender string,
  greeting string,
  guid string,
  index int,
  isactive boolean,
  latitude double,
  longitude double,
  name string,
  phone string,
  picture string,
  registered string,
  tags array<string>)
ROW FORMAT SERDE 'org.openx.data.jsonserde.JsonSerDe'
WITH SERDEPROPERTIES ( "mapping._id" = "id" )
```

The table has a few interesting properties. First off, it is very different than the single column table we used in the UDF example. This means we will be able to select individual rows in a single Hive statement more easily. We also have some complex mappings such as struct and array. These are useful for representing nested structures in the JSON document. Toward the end, we reference the SerDe we added prior to executing our DDL. Finally, we added a SERDEPROPERTIES command. This may not be necessary for all JSON documents but it is necessary for ours because our first column has an illegal underscore. The SERDEPROPERTIES command tells Hive to map the illegal ID to a legal ID, which will prevent an error from occurring.

---

■ **Tip**   Some JSON files can be exceptionally long and complicated. This can make creating the table structure challenging. Luckily there is help. Michael Peterson created a program that will infer a schema based on your JSON file. You can download the code from his GitHub page at `https://github.com/quux00/hive-json-schema`.

---

We can now load data into the table just like we loaded data in the previous UDF example:

```
LOAD DATA INPATH '/tmp/json_data/json1' INTO TABLE json_serde_table;
```

Execute the following query to get some data:

```
SELECT address, friends.name FROM json_serde_table;
```

Notice how we simply use dot notation to access the name value in the friend array. This is an easy and sensible method for accessing nested data.

Another method that many prefer is to use the built-in JSON SerDe for Hive. The steps are similar to the GitHub version except you do not need to add the JAR prior to creating the table. Also if you leave the ID in the original JSON file, the ID will query as NULL. Execute the following DDL to create the table:

```
CREATE TABLE json_serde_table (
  id string,
  about string,
  address string,
  age int,
  balance string,
  company string,
  email string,
  eyecolor string,
  favoritefruit string,
  friends array<struct<id:int, name:string>>,
  gender string,
  greeting string,
  guid string,
  index int,
  isactive boolean,
  latitude double,
  longitude double,
  name string,
  phone string,
  picture string,
  registered string,
  tags array<string>)
ROW FORMAT SERDE 'org.apache.hive.hcatalog.data.JsonSerDe'
STORED AS TEXTFILE;
```

Accessing the native SerDe table is exactly the same as the previous example.

We have looked at two means of accessing JSON data in Hive. This is not an exhaustive list but using a SerDe or the UDF demonstrates the most common and easiest methods of accessing JSON data. JSON is an incredibly functional and common data format and Hive provides an easy means of accessing the data and quickly deriving useful insight from its contents.

# CHAPTER 8

# Hive Analytics

Analytics is the scientific procedure of transforming data into understanding by implementing value-added decisions. So what is Hive analytics? Hive analytics is the practical use of the Hive system to achieve business value.

The objectives of this chapter are to:

- Understand the fundamental building blocks of Hive analytics.

- Understand the fundamental business design tools.

- Create a data warehouse using Hive.

- Combine the fundamental building blocks to succeed with analytics processing.

---

To achieve the maximum learning experience, the reader should complete the chapter's examples in the order they are presented, as the chapter in total forms the analytics structures that should be developed for proficiency in the essential processing skills.

---

## Building an Analytic Model

An analytic model is the base structure for executing queries in order to translate data into knowledge. Wisdom is attained by formulating data structures that will act as the source of the business processes' decisions.

---

■ **Note**    Analytics without effect are wasteful!

---

### Getting Requirements Using Sun Models

The requirements are articulated by communicating a set of sun models. Find out what you plan to achieve with your analytics.

---

■ **Note**    Plan, plan, and then execute. Spend over 80% of your time on the design and then start!

---

© Scott Shaw, Andreas François Vermeulen, Ankur Gupta, David Kjerrumgaard 2016
S. Shaw et al., *Practical Hive*, DOI 10.1007/978-1-4842-0271-5_8

# Business Sun Models

Business sun models are graphical representations of the business query requirements.

The business analyst collects the entire set of analytic requirements from the critical business processes and transforms it from the graphical formats of reports into sun models ready to be developed into Hive code.

By studying each reporting requirement independently, a sun model can be formulated that represents the analysis structure needed to answer the specific report requirement in the Hive data warehouse.

---

■ **Tip**    Keep it simple and study only one specific report at a time.

---

Let's start with a simple model.

## Bar Graph

Let's look at a bar graph as an example of how to deal with the requirements (see Figure 8-1).

What can you extract from the bar graph?

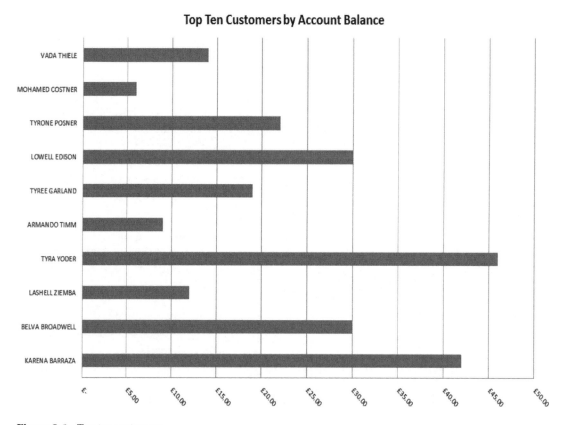

**Figure 8-1.** *Top ten customers*

- Need a selector for `Customer`
- Selector is built from two components:
  - First Name
  - Last Name

- Need a measure for `Balance`
  - The balance is in pounds sterling.
- Need to filter to return "Top Ten Balances".
- Need to order by descending balance.

## Bar Graph with Drop Selections

The graph is enhanced by adding the preference to look after a series of filters to sub-divide the data set (see Figure 8-2).

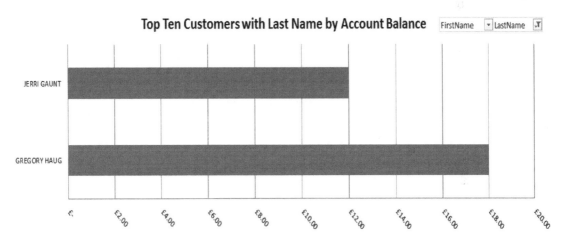

***Figure 8-2.*** *Top ten customers with drop-down lists (FirstName, LastName)*

## Sun Model

The sun model is a business-friendly design tool that enables the business analyst to record the requirements in a format that the business and the technical staff will understand.

A sun model consists of two fundamental components:

- *Dimensions*—The dimensions are the components of the model that can be used to query the analytic model. There are two dimensions in this sun model—Customer and Account (see Figure 8-3).

- *Facts*—The fact is the component of the model that can be used for statistical functions (see Figure 8-3):

  - Sum—Adds the balances of the selected records.

  - Average—Average the balances across the selected records.

  - Maximum—Returns the biggest balance from the selected records.

  - Minimum—Returns the smallest balance from the selected records.

  - Pearson coefficient of correlation of two sets of balances from the selected records.

  - Nth percentile of balances from the set of selected records.

  - Computes a histogram of the balances across the selected records.

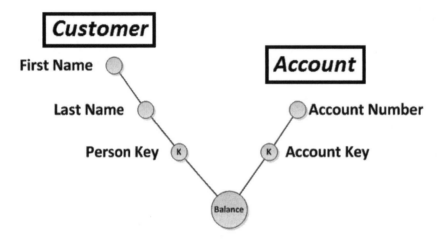

***Figure 8-3.*** *Sun model (two dimensions with one fact)*

Here is a quick abridged explanation of the sun model:

- Left selection is for `Customer`

  - Consists of unique key called `Person Key`

  - Selector for `Last Name`

  - Selector for `First Name`

- Right selection is for `Account`

  - Consists of a unique key called `Account Key`

  - Selector for `Account Number`

At the interaction between `Customer` and `Account`, you record the current `Balance` measure.
Now we examine another relationship, called `LiveAt` (see Figure 8-4), in the business requirements.

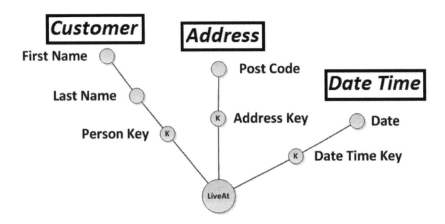

***Figure 8-4.*** *Sun model (three dimensions with one fact)*

- Left selection is for Customer
  - Consists of a unique key called Person Key
  - Selector for Last Name
  - Selector for First Name

- Center selection is for Address
  - Consists of a unique key called Address Key
  - Selector for Post Code

- Right selection is for Date Time
  - Consists of a unique key called Date Time Key
  - Selector for Date

At the interaction between Customer, Address, and Date Time, you record the current LiveAt relationship.

## Interlink Matrix

The interlink matrix is a design tool that assists the business analyst in recording the relationship between dimensions and the facts recorded against each model.

You must create a matrix with the entire unique list of dimensions down the left side of the matrix and the entire list of facts across the top. Then record an indicator in each intersection of the matrix where a specific dimension and specific fact work together to formulate a relationship in the data.

Figure 8-5 shows an example of a matrix that clearly shows that Person is used by both sun models; this shows that Person is a common dimension.

|          | Balance | Live At |
|----------|---------|---------|
| **Time**    |         | X       |
| **Person**  | X       | X       |
| **Account** | X       |         |
| **Address** |         | X       |

***Figure 8-5.*** *Interlink matrix*

---

■ **Note**    As a general rule, it is good practice not to use more than 15 dimensions against a single measure to ensure good performance on queries against the database structure. This reduces the amount of joins during queries by reducing the dimensions per measure.

---

The general rule is the dimensions are "wide" structures, i.e., there will be many selectors. There is a possibility of hundreds of selectors. The records count is "shallow," i.e., there aren't many records, as it has to work as a list in a selector. A dimension has hundreds of entries.

The general rule is the facts are "narrow" structures, i.e., there will be 1 to 15 keys, plus 1 measure. The records count is "deep," i.e., huge volumes of records, as it captures each interaction of the fact in the business. A fact could have billions of entries.

The interlink matrix must be streamlined or cleaned up to ensure an optimal solution. The matrix is formulated by placing the entire list of dimensions down the left side of the matrix. Sort them alphabetically and eliminate any duplicates by transferring the indicators to a single dimension row in the matrix. This action delivers you with your common dimensions.

The top row of the matrix is all the facts or measures you are building for the analytic model. Sort them alphabetically and eliminate any duplicates by transferring the indicators onto a single fact column in the matrix. This action delivers you with your common facts.

## Converting Sun Models to Star Schemas

The set of sun models is converted into a set of star models by adding the technical detail needed to create the physical model. The technique is to take the sun model and add a field type description for each of the selectors and measures.

From the sun model (see Figure 8-4), the First Name now evolves into First Name (varchar (200)), as shown in Figure 8-6.

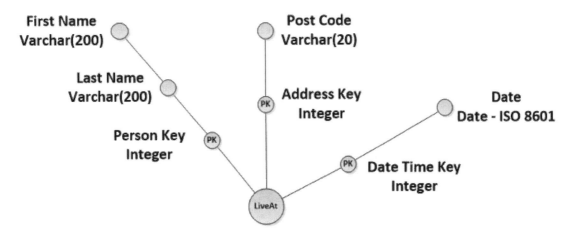

**Figure 8-6.** *Star schema from the sun model*

## Dimensions

Now that you have the basic dimension structures, let's look at the dimensions is more detail.

## Fundamental Dimensions

Dimensions are the part of the data warehouse that enable the "dicing-and slicing: of the data. They are used to subdivide the data set into the required selections.

### Common Types of Dimensions

There is a group of specific types of dimensions you can formulate your data into for your data warehouse model.

Each of the types of dimensions adds specific behaviors into the dimensions and enables the selectors to perform the required business requirements for the analytic model.

The different structures are described as *Types*.

---

■ **Note**    There is a variety of discussion in the design community about which types should exist. We will only cover Type 0, Type 1, Type 2, and Type 3, plus some special other structures that are performance enhancers.

---

So let's discuss these different dimension types in more detail.

---

■ **Tip**    Getting your dimensions spot-on and efficient will take practice, but you can master the process by repeating it until you learn to intuit what works for which types of data.

---

## Type 0: Protect the First Value

The Type 0 dimension record adds a new value only if it does not exist in the dimension table; if it exists, it's kept as the original value that was added to its fields.

This dimension is used when you want to keep the value of the record the same as the first time you received it without any future updates. In businesses, this is used when the original value of the business entity should be protected.

Figure 8-7 shows an example that explains when the first post code of a person is protected.

**Figure 8-7.** *Type 0 dimension*

Ruff Hond is loaded with the KA12 8RR post code during the first run.

Ruff Hond moves to post code EH1 2NG, but the system does not change the post code. It keeps it as KA12 8RR.

## Type 1: Keep Last Value

The Type 1 dimension adds a new value if it does not exist in the dimension table; if it exists it's updated to the latest value.

This dimension is used when you want to keep the value of the record up to date with latest values without keeping any previous values.

The end result of a load is a snapshot of the data as of the last upload.

In businesses, this is used when the last value of the business entity is stored without keeping any history of previous values.

Figure 8-8 shows when the latest post code of a person is stored without history.

| Key | Name | Post Code |
|---|---|---|
| 1 | Ruff Hond | KA12 8RR |
| 2 | Robbie Rot | FK8 1EJ |

First Load

↓

| Key | Name | Post Code |
|---|---|---|
| 1 | Ruff Hond | KA12 8RR |
| 2 | Robbie Rot | FK8 1EJ |

| Key | Name | Post Code |
|---|---|---|
| 1 | Ruff Hond | EH1 2NG |

Second Load

↓

| Key | Name | Post Code |
|---|---|---|
| 1 | Ruff Hond | EH1 2NG |
| 2 | Robbie Rot | FK8 1EJ |

***Figure 8-8.*** *Type 1 dimension*

Ruff Hond is loaded with the KA12 8RR post code during the first run.

Ruff Hond moves to EH1 2NG. The system changes the post code to EH1 2NG and keeps no record of Ruff Hond living at post code KA12 8RR.

## Type 2: Keep Full History

The Type 2 dimension adds a new value if it does not exist in the dimension table; if it exists, the previous current records are updated with a valid date/time value. A new record is added for the latest value.

This dimension is used when you want to keep all the values of the record as the data changes during the lifecycle. This gives you a full history of the data uploads. This is used when the last value of the business entity is stored while keeping a history of all previous values.

Figure 8-9 shows the last post code of a person stored with a full history and the date value being valid.

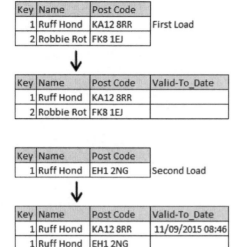

| Key | Name | Post Code |
|---|---|---|
| 1 | Ruff Hond | KA12 8RR | First Load |
| 2 | Robbie Rot | FK8 1EJ |

↓

| Key | Name | Post Code | Valid-To_Date |
|---|---|---|---|
| 1 | Ruff Hond | KA12 8RR | |
| 2 | Robbie Rot | FK8 1EJ | |

| Key | Name | Post Code |
|---|---|---|
| 1 | Ruff Hond | EH1 2NG | Second Load |

↓

| Key | Name | Post Code | Valid-To_Date |
|---|---|---|---|
| 1 | Ruff Hond | KA12 8RR | 11/09/2015 08:46 |
| 1 | Ruff Hond | EH1 2NG | |
| 2 | Robbie Rot | FK8 1EJ | |

***Figure 8-9.*** *Type 2 dimension*

Ruff Hond is loaded with the KA12 8RR post code during the first run with an empty valid-to-date.

Ruff Hond moves to EH1 2NG. The system changes his post code to EH1 2NG by updating the valid-to-date on the previous post code as the KA12 8RR record and then adds a new record with the EH1 2NG post code, with an empty valid-to-date.

---

■ **Note** From experience, we advise that if you are not sure which dimension you need to use, use a Type 2. You can convert the Type 2 dimension to any other later because it holds every data item needed to restructure the Type 2 dimension into any other dimension type.

---

## Type 3: Record Transition

The Type 3 dimension record adds a value if it does not exist in the dimension table; if it exists, the previous field value is updated with the current field value and current field is updated with the last value using the existing data field.

This dimension is used when you want to keep the previous value of the record as the data changes during the lifecycle. This gives you a direct reference to the previous history of the data uploads. This is used when the last value of the business entity is stored while keeping the transition from the previous value in the same record.

Figure 8-10 explains when the last post code of a person is stored with the previous post code.

| Key | Name | Post Code | |
|---|---|---|---|
| 1 | Ruff Hond | KA12 8RR | First Load |
| 2 | Robbie Rot | FK8 1EJ | |

↓

| Key | Name | Post Code | Prev Post Code |
|---|---|---|---|
| 1 | Ruff Hond | KA12 8RR | |
| 2 | Robbie Rot | FK8 1EJ | |

| Key | Name | Post Code | |
|---|---|---|---|
| 1 | Ruff Hond | EH1 2NG | Second Load |

↓

| Key | Name | Post Code | Prev Post Code |
|---|---|---|---|
| 1 | Ruff Hond | EH1 2NG | KA12 8RR |
| 2 | Robbie Rot | FK8 1EJ | |

**Figure 8-10.** *Type 3 dimension*

Ruff Hond is loaded with post code KA12 8RR during the first run, with an empty Prev Post Code value. Ruff Hond moves to EH1 2NG. The system changes the post code to EH1 2NG by updating the Prev Post Code to KA12 8RR on the record and then updating a record with a post code to EH1 2NG.

## Mini-Dimensions

The mini-dimension (see Figure 8-11) is an extension of the dimension. It supports the subdivision of the dimension to assist with data query fields in these cases:

- When fields are not used by all query processes in the model.

- When it is not possible to return all fields in one dimension during one query action.

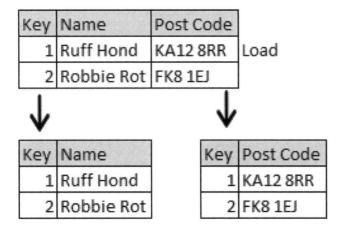

**Figure 8-11.** *Mini-dimensions*

Dividing the fields into two dimensions reduces the size of the data the queries have to process. This works well if you only need part of the field on a regular basis. This does not change the values of the record; it only enhances the processing speed of the query.

## Mini-Dimension for Fast-Changing Values in Type 2 Dimensions

If the dimension includes specific fields that undergo fast changes that result in the dimension growing too fast on the disk, you can split off these fast-changing fields (see Figure 8-12) to remodel the data warehouse. You can minimize the disk size growth in this manner with ease.

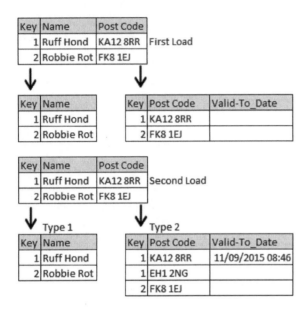

*Figure 8-12. Fast-changing the mini-dimension*

Dividing the fields into two dimensions reduces the size of the data stored to keep the history of the data records.

This does not change the values of the record; it only improves the data storage and query process.

---

■ **Note**    There is a fine balance between disk space growth and query time impact is required. You will be advised to tune these structures over time to maintain good performance.

---

## Mini-Dimension for Separated Values Due to Security Constraints

It's often mandatory to isolate values to ensure compliance with security requirements. In this structure, you separate the security-sensitive fields in a detached dimension. See Figure 8-13.

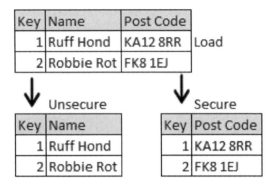

*Figure 8-13.* *Security mini-dimensions*

Dividing the fields into two dimensions isolates the fields across two dimensions to enable the data to be secured in isolation. (See Chapter 9 for how to use security in Hive.)

---

■ **Caution**    Ensure that you keep the keys in synchronization across the complete data set. If you lose these relationships, your whole structure will become null and void.

---

## Mini-Dimension for Separated Values Due to Language Differences

There is the requirement in analytic models to present the same dimension in different languages. This is achieved by replication of the dimensions with each language in a separate dimension. This makes the process easier than using one big dimension with all the languages values. See Figure 8-14.

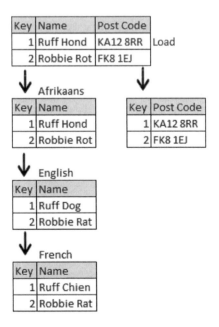

*Figure 8-14.* *Mini-dimension for language differences*

By dividing the fields into many mini-dimensions, queries can generate different languages. You simply combine the correct data queries tables for the language you require.

## Outrigger Dimensions

An outrigger is used when you already have a dimension that contains the value you want, so you simply add a key into the dimension you construct to link to the existing dimension. See Figure 8-15.

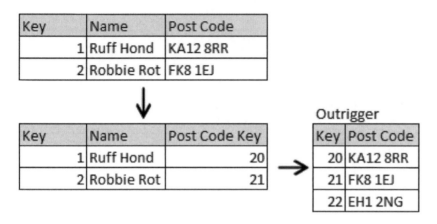

***Figure 8-15.*** *Outrigger dimensions*

This creates three tables that are used to represent the required data.

---

■ **Caution**    If you create an outrigger, take care that during current and future changes you do not implement changes to the outrigger structure. That would damage the main purpose of the dimension you used as the outrigger, as it would null and void the outrigger relationship and the main purpose of the dimension.

The most common mistake is using an automatic key generator on the outrigger dimension. Every time your rebuild the key you disrupt the outrigger relationship.

---

## Bridge Dimensions

The bridge dimension is used to represent the relationship created when two dimensions have a many-to-many relationship and you want to create a bridge dimension structure. See Figure 8-16.

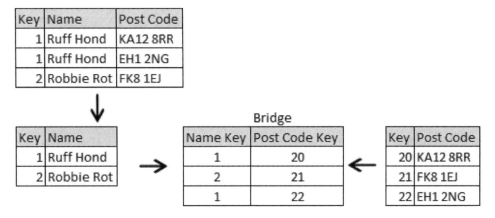

***Figure 8-16.*** *Bridge dimensions*

Ruff Hond lives at two locations—KA12 8RR and EH1 2NG. By adding the bridge, this relationship is converted into two one-to-many relationships.

___

■ **Caution**    If you create a bridge dimension, it should be kept to a minimum, as these structures require complex relationships when you query them. These data structures can create massive data sets during queries.

___

# Facts

Facts are the measures of the analytic model. The data fields are numeric to enable the option to apply mathematical and aggregation functions.

## Calculated Facts

Using mathematical and aggregation functions creates new facts. Possible functions to use are:

- Sum
- Average
- Minimum
- Maximum
- Count
- Combining facts to create a new calculated fact

There are many other functions you can apply, but we are not going list them here. (See Appendix B for more details.)

Figure 8-17 shows how applying a sum creates a new calculated fact called the current balance.

| Key | Name | Post Code | Transaction |
|-----|------|-----------|-------------|
| 1 | Ruff Hond | KA12 8RR | £ 100.60 |
| 2 | Ruff Hond | KA12 8RR | £ 20.00 |
| 3 | Ruff Hond | KA12 8RR | -£ 10.00 |
| 4 | Robbie Rot | FK8 1EJ | £ 200.00 |
| 5 | Robbie Rot | FK8 1EJ | -£ 30.00 |

↓

| Key | Name | Post Code | Balance |
|-----|------|-----------|---------|
| 1 | Ruff Hond | KA12 8RR | £ 110.60 |
| 2 | Robbie Rot | FK8 1EJ | £ 170.00 |

*Figure 8-17. Calculated facts*

## Factless Facts

A factless fact is a data structure that presents a structure that holds the relationship between the different dimensions.

There are relationships between entities that have no measures. An example is the relationship between customers and their home addresses (see Figure 8-18).

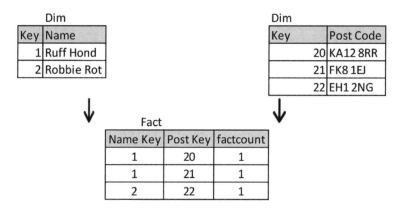

*Figure 8-18. Factless fact*

There is only the relationship between the customer and the address in this fact table. So no fact or measure is stored with the keys.

---

■ **Note** Word of advice, when creating facts, you should always add a standard field called factcount = 1, as this makes it easier to use mathematical and aggregation functions on queries.

---

# Building the Data Warehouse

The data warehouse is built by converting sun models into star models. You do this by providing fields with data types and then translating the star schemas into Hive code to build the Hive data warehouse structures.

Before you proceed with the physical construction, let's just do a validation check:

Revisit the interlink matrix and all the sun models you prepared with the business.

- The matrix is formulated by placing the dimensions down the left side of the matrix. Sort them alphabetically and remove any duplicates. Now you have your common dimensions.

- The top row of the matrix is all the facts and measures you are building for the analytic model. Sort them alphabetically and remove any duplicates. Now you have your common facts.

- Determine if the dimensions have the correct type your business needs.

    - Type 0

    - Type 1

    - Type 2

    - Type 3

    - Mini

        - Fast-Changing

        - Security

        - Language

    - Outriggers

    - Bridges

Now that you validated the data warehouse, let's create a dimension in Hive.

---

■ **Note**    To execute the Hive code, you need to open your Hive terminal.

---

## Log On as the Root User

If you receive an access error against the root user, execute following commands:

```
hadoop fs -mkdir /user/root
hadoop fs -chmod 777 /user/root
```

This will resolve the access issue.

## Dimensions

Dimensions are the core selectors of the data warehouse. Dimensions are created by using tables with the prefix of dim-.

## Typical Dimension

This is a typical dimension structure.

- It has a unique key called personkey
- It has two selectors called firstname and lastname

There are two key pieces needed to build a simple Hive dimension: a database and a table.

1. To create a transform database, execute the following in your Hive terminal:

```
CREATE DATABASE IF NOT EXISTS transformdb;
```

This Hive code creates a database called transformdb while checking that it did not exists.

---

■ **Note**    If you are wondering, why transformdb, this will be discussed in detail later in the chapter in the section "Master Data Warehouse Management". Just use it as specified until that point.

---

2. Create a person dimension table.

The dimension consists of:

- Personkey, which is the key of the dimension.
- Firstname and lastname, which are the attributes of the dimension.

In your Hive terminal, execute the following:

```
USE transformdb;
```

This informs Hive to use the database you just created.
In your Hive terminal, execute the following to create the dimension table:

```
CREATE TABLE IF NOT EXISTS transformdb.dimperson (
  personkey  BIGINT,
  firstname  STRING,
  lastname   STRING
)
CLUSTERED BY (firstname, lastname,personkey) INTO 1 BUCKETS
STORED AS orc
TBLPROPERTIES('transactional' = 'true','orc.compress'='ZLIB','orc.create.index'='true');
```

This Hive code creates a table called transformdb.dimperson with three fields.

---

■ **Note**    If you are unsure about the full meaning of the Hive command, read Chapter 4 for more information.

---

## Common Dimensions

The common dimensions are the communal selectors you require for your analytic model. This is the base for all your possible drop lists and/or filters you can apply to the model.

At this point you will not create the rest of the dimensions, as they are created in the "Master Data Warehouse Management" section, in the "Transform Database" subheading.

# Facts

Facts are the measures of the analytic model.

For facts, you create the tables with the prefix of `fct-`.

## Typical Facts

The following is a typical fact structure:

- A set of keys, one for every dimension linked to the fact.

- A single fact, i.e., a measure.

To create the fact table, execute the following in your Hive terminal:

```
CREATE TABLE IF NOT EXISTS transformdb.fctpersonaccount (
  personaccountkey    BIGINT,
  personkey           BIGINT,
  accountkey          BIGINT,
  balance             DECIMAL(18, 9)
)
CLUSTERED BY (personkey,accountkey) INTO 1 BUCKETS
STORED AS orc
TBLPROPERTIES('transactional' = 'true','orc.compress'='ZLIB','orc.create.index'='true');
```

This Hive code creates a table called `fctpersonaccount` that consists of keys that link to three dimensions (`personaccount`, `person`, and `account`) and has one fact called `balance`.

## Common Facts

The common facts are the common measures you require for your analytic model. The fact can be used with any Hive mathematical and aggregate functions.

■ **Note**    See Appendix B for details on what you can use.

There are the various possibilities to create extra calculated facts at query time to supplement these common facts.

Examples include:

- Final balance—Use a sum() function.

- Maximum balance—Use a max() function.

- Amount of accounts—Use a count() function.

- Variance in balance—Use a variance() function.

- Percentile of balance—Use a percentile_approx() function.

# Assessing an Analytic Model

You have now constructed a basic analytic model. The next step is to enable the queries to assess the model for the users to achieve their business requirements.

## Assess the Sun Models

A good test is to take each of your sun models and create a query that delivers the information in the required format to the business community.

This way, you create a one-to-one delivery check against the agreed sun models that you formulated with your business user's help.

In your Hive terminal, execute the following to create the extra Hive structures you will need.

This should be easy as you will have mastered the required Hive skills by now.

## Create Two More Databases

You will need extra databases and tables for the next step. For now, simply create them; the business explanation is covered in the "Master Data Warehouse Management" section.

```
CREATE DATABASE IF NOT EXISTS organisedb;
CREATE DATABASE IF NOT EXISTS reportdb;
```

## Create Extra Tables

Now let's add more tables to give you extra structures to master your Hive skills even more. Practice makes perfect, so follow these eight steps.

1. In database, use transformdb:

```
USE transformdb;
```

2. Create the table transformdb.dimaccount:

```
CREATE TABLE IF NOT EXISTS transformdb.dimaccount (
    accountkey      BIGINT,
    accountnumber   INT
)
```

```
CLUSTERED BY (accountnumber,accountkey) INTO 1 BUCKETS
STORED AS orc
TBLPROPERTIES('transactional' = 'true','orc.compress'='ZLIB','orc.create.index'='true');
```

3. In database, use organisedb:

```
USE organisedb;
```

4. Create the table organisedb.dimaccount:

```
CREATE TABLE IF NOT EXISTS organisedb.dimaccount LIKE transformdb.dimaccount;
```

---

Did you spot the use of like to create new tables? This is a useful command to ensure your structures match between two tables.

---

5. Create the table organisedb.fctpersonaccount:

```
CREATE TABLE IF NOT EXISTS organisedb.fctpersonaccount (
  personaccountkey      BIGINT,
  personkey             BIGINT,
  accountkey            BIGINT,
  balance               DECIMAL(18, 9)
)
CLUSTERED BY (personkey,accountkey) INTO 1 BUCKETS
STORED AS orc
TBLPROPERTIES('transactional' = true','orc.compress'='ZLIB', 'orc.create.index'='true');
```

6. Create the table organisedb.dimperson:

```
CREATE TABLE IF NOT EXISTS organisedb.dimperson (
  personkey  BIGINT,
  firstname  STRING,
  lastname   STRING
)
CLUSTERED BY (firstname, lastname,personkey) INTO 1 BUCKETS
STORED AS orc
TBLPROPERTIES('transactional' = 'true','orc.compress'='ZLIB','orc.create.index'='true');
```

7. In database, use reportdb:

```
USE reportdb;
```

8. Create the table reportdb.report001:

```
CREATE TABLE IF NOT EXISTS reportdb.report001(
  firstname       STRING,
  lastname        STRING,
  accountnumber   INT,
  balance         DECIMAL(18, 9)
)
CLUSTERED BY (firstname, lastname) INTO 1 BUCKETS
STORED AS orc
TBLPROPERTIES('transactional' = 'true','orc.compress'='ZLIB','orc.create.index'='true');
```

You now have all the data structures you will require for our next steps.

The assessment is to test if you have the complete sun model (see Figure 8-19).

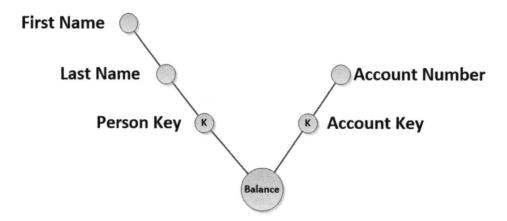

***Figure 8-19.*** *Sun model*

To create the assessment for the sun model (see Figure 8-19), use the following Hive code.

```
INSERT INTO TABLE reportdb.report001
SELECT
  dimperson.firstname, dimperson.lastname,
  dimaccount.accountnumber,
  fctpersonaccount.balance
FROM
  organisedb.fctpersonaccount
JOIN
  organisedb.dimperson
ON
  fctpersonaccount.personkey = dimperson.personkey
JOIN
  organisedb.dimaccount
ON
  fctpersonaccount.accountkey = dimaccount.accountkey;
```

If you successfully return a result you have proven that the sun model was produced by your data warehouse structures.

## Assess the Aggregations

Creating aggregations against the analytic model is common and is covered by the calculated fact structures.

There is also the option to apply some complex functional calculations, but for the purposes of this structure, you will only note the sum option.

Create the table reportdb.report002 as follows:

```
CREATE TABLE IF NOT EXISTS reportdb.report002(
  accountnumber    INT,
  last_balance          DECIMAL(18, 9)
)
CLUSTERED BY (firstname, lastname) INTO 1 BUCKETS
STORED AS orc
TBLPROPERTIES('transactional' = 'true','orc.compress'='ZLIB','orc.create.index'='true');
```

Aggregate the data:

```
INSERT INTO TABLE reportdb.report002
SELECT
  dimaccount.accountnumber,
  sum(fctpersonaccount.balance) as last_balance
FROM
  organisedb.fctpersonaccount
JOIN
  organisedb.dimaccount
ON
  fctpersonaccount.accountkey = dimaccount.accountkey;
```

This Hive code aggregates the balances to give the latest balance measure by using the sum() function.

## Assess the Data Marts

The concept of a data mart is used when you need to subdivide into smaller analytic models for a specific analytic requirement that is permanently stored.

Possible reasons you would need a data mart include:

- To divide data by region, so that each region only sees its own data.

- To create month-end results that are static until the next month end.

- To enhance performance by reducing data volumes per query.

- To subdivide the data warehouse to have data marts to place physically on servers in the branches.

To perform this assessment, execute the following Hive code:

```
INSERT INTO TABLE organisedb.fctpersonaccount
SELECT DISTINCT
  personaccountkey,
  personkey,
  accountkey,
  balance
FROM
  transformdb.fctpersonaccount
WHERE
  personaccountkey = 1
ORDER BY personaccountkey,personkey,accountkey;
```

This code performs a subdivide on the table `transformdb.fctpersonaccount` and inserts only the records that match the `where` statement `personaccountkey = 1` into table "`organisedb.fctpersonaccount`".

This can be used to create data marts for specific subsets of the data warehouse.

You now understand the basic processes of a working data warehouse, so let's build a full data warehouse.

# Master Data Warehouse Management

Now that we have explained the theories behind building a data warehouse model by creating a data warehouse and data marts, the following several examples show a complete cycle of the build process for a simple set of requirements.

We are using the ***Retrieve-Assess-Process-Transform-Organize-Report*** design principle from the Rapid Information Factory approach.

As you go through the complete data warehouse examples, we will discuss what each layer of the design supplies to the data warehouse build process.

---

■ **Note**    The next part of the chapter is a full processing of the warehouse. If you can complete the remainder of this chapter, you have mastered the data warehouse in Hive. You will find the Hive code in our example section to assist you with the process, which saves you from coping it from the book.

---

A data warehouse is a structure with layers and as a unit will enable you to handle your business requirements (your sun models).

Remember the following advice we learned through experience:

- Plan every change to the data warehouse structure with care and you will be successful.

- Use the sun models to verify your requirements with the business.

- Keep to the processing rules!

- Taking shortcuts will cost you in the future.

# Prerequisites

You need the example data from the 00rawdata directory. The following comma-separated value files are required:

- The rawaccount.csv file—Holds 10,000 records.

- The rawaddress.csv file—Holds 220,182 records.

- The rawaddresshistory.csv file—Holds 100 records.

- The rawdatetime.csv file—Holds 1,052,640 records

- The rawfirstname.csv file—Holds 5,494 records.

- The rawlastname.csv file—Holds 16,001 records

- The rawperson.csv file—Holds 1,000 records.

Now that you have your data loaded, let's build the data warehouse.

# Retrieve Database

The retrieve database is the data area that is used to transfer the data from the external data sources into the Hive data structure.

Data is normally transferred into the structure in an as-is format. Simply replicate the data structure and the data contained within the structure from the external data source.

---

## WHY AS-IS ... ?

This enables you to reprocess your data warehouse from the original data format without dependencies on other source systems. We have learned the hard way that a reformatting process does not always work as designed. So keep the original data in Hive; it's less painful that way.

---

You create a database called retrievedb to hold the imported data.

---

■ **Note**    Source code for this chapter is available for download from www.apress.com/9781484202722. See example script Retrieve001.txt for the Hive code.

---

Let's start by removing the existing retrievedb. Remember you created this earlier.

```
DROP DATABASE retrievedb CASCADE;
```

Now you recreate retrievedb to accept the data from your external data sources.

```
CREATE DATABASE IF NOT EXISTS retrievedb;
```

You will now create tables and load data from the external data sources.
Simply follow the steps to load the required set of data.

1. Create the table and load the data for rawfirstname.csv:

```
USE retrievedb;
CREATE TABLE IF NOT EXISTS retrievedb.rawfirstname (
  firstnameid    string,
  firstname      string,
  sex            string
)
ROW FORMAT DELIMITED FIELDS TERMINATED BY ',';

LOAD DATA LOCAL INPATH 'file:///root/exampledata/datawarehouse/00rawdata/rawfirstname.csv'
OVERWRITE INTO TABLE retrievedb.rawfirstname;
```

2. Create the table and load the data for rawlastname.csv:

```
CREATE TABLE IF NOT EXISTS retrievedb.rawlastname (
  lastnameid     string,
  lastname       string
)
ROW FORMAT DELIMITED FIELDS TERMINATED BY ',';

LOAD DATA LOCAL INPATH 'file:///root/exampledata/datawarehouse/00rawdata/rawlastname.csv'
OVERWRITE INTO TABLE retrievedb.rawlastname;
```

3. Create the table and load the data for rawperson.csv:

```
CREATE TABLE IF NOT EXISTS retrievedb.rawperson (
  persid         string,
  firstnameid    string,
  lastnameid     string
)
ROW FORMAT DELIMITED FIELDS TERMINATED BY ',';

LOAD DATA LOCAL INPATH 'file:///root/exampledata/datawarehouse/00rawdata/rawperson.csv'
OVERWRITE INTO TABLE retrievedb.rawperson;
```

## Additional Data Loads

Let's do some more data loads. (See example script Retrieve002.txt for the Hive code.)

1. Create the table and load the data for rawdatetime.csv:

```
CREATE TABLE IF NOT EXISTS retrievedb.rawdatetime (
  id             string, datetimes      string, monthname  string,
  yearnumber     string, monthnumber    string, daynumber  string,
  hournumber     string, minutenumber   string, ampm       string
)
ROW FORMAT DELIMITED FIELDS TERMINATED BY ',';

LOAD DATA LOCAL INPATH 'file:///root/exampledata/datawarehouse/00rawdata/rawdatetime.csv'
OVERWRITE INTO TABLE retrievedb.rawdatetime;
```

See example script `Retrieve003.txt` for the Hive code.

2.   Create the table and load the data for `rawaddress.csv`:

```
CREATE TABLE IF NOT EXISTS retrievedb.rawaddress (
   id            string, Postcode       string, Latitude string,
   Longitude     string, Easting        string, Northing string,
   GridRef       string, District       string, Ward     string,
   DistrictCode  string, WardCode       string, Country  string,
   CountyCode    string, Constituency   string, TypeArea string
)
ROW FORMAT DELIMITED FIELDS TERMINATED BY ',';

LOAD DATA LOCAL INPATH 'file:///root/exampledata/datawarehouse/00rawdata/rawaddress.csv'
OVERWRITE INTO TABLE retrievedb.rawaddress;
```

3.   Create the table and load the data for `rawaddresshistory.csv`:

```
CREATE TABLE IF NOT EXISTS retrievedb.rawaddresshistory (
   id string, pid  string, aid  string, did1  string, did2  string
)
ROW FORMAT DELIMITED FIELDS TERMINATED BY ',';

LOAD DATA LOCAL INPATH 'file:///root/exampledata/datawarehouse/00rawdata/rawaddresshistory.
csv' OVERWRITE INTO TABLE retrievedb.rawaddresshistory;
```

See example script `Retrieve004.txt` for the Hive code.

4.   Create the table and load the data for `rawaccount.csv`:

```
CREATE TABLE IF NOT EXISTS retrievedb.rawaccount (
   id    string,  pid string,  accountno  string, balance  string
)
ROW FORMAT DELIMITED FIELDS TERMINATED BY ',';

LOAD DATA LOCAL INPATH 'file:///root/exampledata/datawarehouse/00rawdata/rawaccount.csv'
OVERWRITE INTO TABLE retrievedb.rawaccount;
```

You have just completed the data retrieve layer of the data warehouse and mastered the following:

- Creating tables with delimited fields.
- Loading data from delimited files.

The same Hive code can also support other delimiters.

Try the pipe delimiter.

```
CREATE TABLE IF NOT EXISTS retrievedb.rawaccountpipe (
   id   string,   pid string,   accountno  string, balance  string
)
ROW FORMAT DELIMITED FIELDS TERMINATED BY '|';

LOAD DATA LOCAL INPATH 'file:///root/exampledata/datawarehouse/00rawdata/rawaccount.pipe'
OVERWRITE INTO TABLE retrievedb.rawaccount;
```

Any delimiter is possible, but comma, tab, pipe, and space are the more common.

## Assess Database

The Assess Database is the data structure that enables you to use data quality rules to assess if the data in your retrieve database is of good quality.

The assess process is basically a process of channeling the data from one table to the next to ensure the specific assessment function is performed.

This results in a series of interim tables that, after the process is completed, are discarded.

■ **Tip**    Suffix your interim tables with a number. For example, firstname001 belongs to firstname's process.

To enable the process, you create a database called assessdb.

See example script Assess001.txt for the Hive code.

## Remove the access Database

```
DROP DATABASE IF EXISTS assessdb CASCADE;
```

## Create the assess Database

```
CREATE DATABASE IF NOT EXISTS assessdb;
```

## Create the assess firstname Tables

```
USE assessdb;
```

The assess layer is now used to assess and clean up the firstname data from the retrieve layer.

## Create the Interim firstname001 Table

```
CREATE TABLE IF NOT EXISTS assessdb.firstname001 (
  firstnameid    string,
  firstname      string,
  sex            string
)
CLUSTERED BY (firstnameid) INTO 1 BUCKETS
STORED AS orc
TBLPROPERTIES('transactional' = 'true','orc.compress'='ZLIB','orc.create.index'='true');
```

1.  Clear out all data from firstname001.

```
TRUNCATE TABLE assessdb.firstname001;
```

## Remove the Headings from the firstname Data

The first assessment is on the firstname.

On investigation of the retrievedb.rawfirstname, we discovered that due to the structure mismatch between the input file and the database, we in error uploaded the headings of the input file.

The proposed solution is to simply filter the headings out of the data set by using a SELECT statement with a WHERE of firstnameid <> '"id"'.

```
INSERT INTO TABLE assessdb.firstname001
SELECT firstnameid, firstname, sex
FROM retrievedb.rawfirstname
WHERE firstnameid <> '"id"';
```

## Create the Interim firstname002 Table

You need to create the table assessdb.firstname002 and then perform the INSERT statement to assess the data.

```
CREATE TABLE IF NOT EXISTS assessdb.firstname002 (
  firstnameid    string,
  firstname      string,
  sex            string
)
CLUSTERED BY (firstnameid) INTO 1 BUCKETS
STORED AS orc
TBLPROPERTIES('transactional' = 'true','orc.compress'='ZLIB','orc.create.index'='true');
```

## Clear Out All Data from firstname002

```
TRUNCATE TABLE assessdb.firstname002;
```

## Remove the Spaces from the Firstname Data

Now that you have the data set without the headings, you can assess the quality of the records.

We discovered because of quality checks in our source system that values in the data set may have leading or lagging spaces.

To fix this issue, we use built-in functions in Hive.

We will use:

- ltrim—Left trim removes any leading spaces.

- rtrim —Right trim removes any lagging spaces.

We also compound the two functions into a function chain by using rtrim(ltrim()).

To complete this assess rule, we create a SELECT statement to apply our new function to the data in firstname001 and then insert that into a table called firstname002.

```
INSERT INTO TABLE assessdb.firstname002
SELECT firstnameid, rtrim(ltrim(firstname)), rtrim(ltrim(sex))
FROM assessdb.firstname001;
```

## Create the Interim firstname003 Table

You need to create the table assessdb.firstname003 and then perform the INSERT statement.

```
CREATE TABLE IF NOT EXISTS assessdb.firstname003 (
   firstnameid    int,
   firstname      string,
   sex            string
)
CLUSTERED BY (firstnameid) INTO 1 BUCKETS
STORED AS orc
TBLPROPERTIES('transactional' = 'true','orc.compress'='ZLIB','orc.create.index'='true');
```

## Clear Out All Data from firstname003

```
TRUNCATE TABLE assessdb.firstname003;
```

## Convert Data Types in the firstname Data

On further inspection of the data set, we discover two more issues:

- We have to convert the firstnameid from string to integer data type.

- We have to remove an unwanted extra character from the firstname and sex data values.

Hive has internal functions to deal with these issues also.

■ **Tip** We suggest you research the internal functions in Appendix B. Understand how they work and try combinations of them in a chain. These are your tools—understand them and master them.

To complete this assess rule, we create a SELECT statement to apply our new functions to the data in firstname002 and then insert that into a table called firstname003.

```
INSERT INTO TABLE assessdb.firstname003
SELECT
  CAST(firstnameid as INT), SUBSTRING(firstname,2,LENGTH(firstname)-2),
  SUBSTRING(sex,2,LENGTH(sex)-2)
FROM assessdb.firstname002;
```

## Create the firstname Table

```
CREATE TABLE IF NOT EXISTS assessdb.firstname (
  firstnameid    int,
  firstname      string,
  sex            string
)
CLUSTERED BY (firstnameid) INTO 1 BUCKETS
STORED AS orc
TBLPROPERTIES('transactional' = 'true','orc.compress'='ZLIB','orc.create.index'='true');
```

## Clear Out All Data from firstname

```
TRUNCATE TABLE assessdb.firstname;
```

## Transfer Data in the firstname Table

Perfect, we now have a high-quality data set in assessdb.firstname003.

You will now transfer the data set to the final assess table.

To complete this assess rule, we create a SELECT statement to our high quality data set from firstname003 and then insert it into a table called firstname.

```
INSERT INTO TABLE assessdb.firstname
SELECT
  firstnameid,
  firstname,
  sex
FROM
  assessdb.firstname003
ORDER BY firstnameid;
```

■ **Tip**    You could be ensnared to simply go back to the previous step and point the `insert` to `firstname` and not `firstname003`. It would be a valid process, but it's better to always clean up the data set first and then load it to the final table. Here is why.

You can perform extra steps in the assessment chain without impacting the existing data set in final table. This helps with future development.

The use of filters and functions will always be slower than a direct `select` and `insert`. So if you prepare the data set first by using the filters and functions and then you simply insert you data, the final table will be unstable for a shorter time between truncating the table and inserting the new data set.

## Evaluate Data in the firstname Table

```
SELECT
  firstnameid,
  firstname,
  sex
from
  assessdb.firstname
SORT BY
  firstname LIMIT 10;
```

## What Have You Mastered?

- You can remove unwanted records, i.e., headings.
- You can remove unwanted spaces in the data records.
- You can change data types of the data set.

You can now apply your new knowledge in the `lastname` tables using the next set of data.

## Create assess lastname Tables

```
CREATE TABLE IF NOT EXISTS assessdb.lastname001 (
  lastnameid     string,
  lastname       string
)
CLUSTERED BY (lastnameid) INTO 1 BUCKETS
STORED AS orc
TBLPROPERTIES('transactional' = 'true','orc.compress'='ZLIB','orc.create.index'='true');

TRUNCATE TABLE assessdb.lastname001;

INSERT INTO TABLE assessdb.lastname001
SELECT lastnameid, lastname
FROM retrievedb.rawlastname
WHERE lastnameid <> '"id"';
```

```
CREATE TABLE IF NOT EXISTS assessdb.lastname002 (
   lastnameid    string,
   lastname      string
)
CLUSTERED BY (lastnameid) INTO 1 BUCKETS
STORED AS orc
TBLPROPERTIES('transactional' = 'true','orc.compress'='ZLIB','orc.create.index'='true');

TRUNCATE TABLE assessdb.lastname002;

INSERT INTO TABLE assessdb.lastname002
SELECT lastnameid, rtrim(ltrim(lastname))
FROM assessdb.lastname001;

CREATE TABLE IF NOT EXISTS assessdb.lastname003 (
   lastnameid    int,
   lastname      string
)
CLUSTERED BY (lastnameid) INTO 1 BUCKETS
STORED AS orc
TBLPROPERTIES('transactional' = 'true','orc.compress'='ZLIB','orc.create.index'='true');

TRUNCATE TABLE assessdb.lastname003;

INSERT INTO TABLE assessdb.lastname003
SELECT CAST(lastnameid as INT), SUBSTRING(lastname,2,LENGTH(lastname)-2)
FROM assessdb.lastname002;

CREATE TABLE IF NOT EXISTS assessdb.lastname (
   lastnameid    int,
   lastname      string
)
CLUSTERED BY (lastnameid) INTO 1 BUCKETS
STORED AS orc
TBLPROPERTIES('transactional' = 'true','orc.compress'='ZLIB','orc.create.index'='true');

TRUNCATE TABLE assessdb.lastname;

INSERT INTO TABLE assessdb.lastname
SELECT lastnameid, lastname
FROM assessdb.lastname003
ORDER BY lastnameid;
```

## Evaluate Data in the lastname Table

```
SELECT
   lastnameid,
   lastname
```

```
from
  assessdb.lastname
SORT BY
  lastname LIMIT 10;
```

If you see 10 records, you've created the next table. Let's move on.

## Create assess person Tables

```
CREATE TABLE IF NOT EXISTS assessdb.person001 (
  persid          string,
  firstnameid     string,
  lastnameid      string
)
CLUSTERED BY (persid) INTO 1 BUCKETS
STORED AS orc
TBLPROPERTIES('transactional' = 'true','orc.compress'='ZLIB','orc.create.index'='true');

TRUNCATE TABLE assessdb.person001;

INSERT INTO TABLE assessdb.person001
SELECT persid, firstnameid, lastnameid
FROM retrievedb.rawperson
WHERE persid <> '"id"';

CREATE TABLE IF NOT EXISTS assessdb.person002 (
  persid          int,
  firstnameid     int,
  lastnameid      int
)
CLUSTERED BY (persid) INTO 1 BUCKETS
STORED AS orc
TBLPROPERTIES('transactional' = 'true','orc.compress'='ZLIB','orc.create.index'='true');

TRUNCATE TABLE assessdb.person002;

INSERT INTO TABLE assessdb.person002
SELECT CAST(persid as INT), CAST(firstnameid as INT), CAST(lastnameid as INT)
FROM assessdb.person001;

CREATE TABLE IF NOT EXISTS assessdb.person (
  persid          int,
  firstnameid     int,
  lastnameid      int
)
CLUSTERED BY (persid) INTO 1 BUCKETS
STORED AS orc
TBLPROPERTIES('transactional' = 'true','orc.compress'='ZLIB','orc.create.index'='true');
```

```
TRUNCATE TABLE assessdb.person;

INSERT INTO TABLE assessdb.person
SELECT persid, firstnameid, lastnameid
FROM assessdb.person002;
```

The next table type is a combination table. Combination tables are formulated form more than one source table.

## Create assess personfull Tables

```
CREATE TABLE IF NOT EXISTS assessdb.personfull(
  persid        int,
  firstnameid   int,
  firstname     string,
  lastnameid    int,
  lastname      string,
  sex           string
)
CLUSTERED BY (persid) INTO 1 BUCKETS
STORED AS orc
TBLPROPERTIES('transactional' = 'true','orc.compress'='ZLIB','orc.create.index'='true');

TRUNCATE TABLE assessdb.personfull;
```

Let's master this combination table type:

```
INSERT INTO TABLE assessdb.personfull
SELECT person.persid, person.firstnameid, firstname.firstname,          person.lastnameid,
lastname.lastname, firstname.sex
FROM assessdb.firstname
JOIN assessdb.person
ON firstname.firstnameid = person.firstnameid
JOIN assessdb.lastname
ON lastname.lastnameid = person.lastnameid;
```

---

■ **Note**    You can now create tables directly from retrieving data and from a combination of other assess tables. You can do combination tables using joins. See Chapter 5 on joins for more detail.

---

## Cleanup assess Database

The next step is tidying up the assess layer. This reclaims extra space for the next steps.

```
DROP TABLE assessdb.firstname001;
DROP TABLE assessdb.firstname002;
DROP TABLE assessdb.firstname003;
DROP TABLE assessdb.lastname001;
```

```
DROP TABLE assessdb.lastname002;
DROP TABLE assessdb.lastname003;
DROP TABLE assessdb.person001;
DROP TABLE assessdb.person002;
```

See example script `Assess002.txt` for the Hive code.

Now that you have mastered the process of assessing data, you can try your skills against a larger data set.

## Create assess datetime Tables

```
CREATE TABLE IF NOT EXISTS assessdb.datetime001 (
  id            string, datetimes      string, monthname string,
  yearnumber    string, monthnumber    string, daynumber string,
  hournumber    string, minutenumber   string, ampm        string
)
CLUSTERED BY (id) INTO 1 BUCKETS
STORED AS orc
TBLPROPERTIES('transactional' = 'true','orc.compress'='ZLIB','orc.create.index'='true');

TRUNCATE TABLE assessdb.datetime001;

INSERT INTO TABLE assessdb.datetime001
SELECT
  id, datetimes, monthname, yearnumber, monthnumber,
  daynumber, hournumber, minutenumber, ampm
FROM retrievedb.rawdatetime
WHERE id <> '"id"';

CREATE TABLE IF NOT EXISTS assessdb.datetime002 (
  id           string, datetimes     string, monthname string,
  yearnumber string, monthnumber   string, daynumber string,
  hournumber string, minutenumber string, ampm        string
)
CLUSTERED BY (id) INTO 1 BUCKETS
STORED AS orc
TBLPROPERTIES('transactional' = 'true','orc.compress'='ZLIB','orc.create.index'='true');

TRUNCATE TABLE assessdb.datetime002;

INSERT INTO TABLE assessdb.datetime002
SELECT
  id, rtrim(ltrim(datetimes)), rtrim(ltrim(monthname)),
  rtrim(ltrim(yearnumber)), rtrim(ltrim(monthnumber)),
  rtrim(ltrim(daynumber)), rtrim(ltrim(hournumber)),
  rtrim(ltrim(minutenumber)), rtrim(ltrim(ampm))
FROM assessdb.datetime001;
```

```
CREATE TABLE IF NOT EXISTS assessdb.datetime003 (
  id            int, datetimes     string, monthname string,
  yearnumber  int, monthnumber  int,    daynumber int,
  hournumber  int, minutenumber int,    ampm       string
)
CLUSTERED BY (id) INTO 1 BUCKETS
STORED AS orc
TBLPROPERTIES('transactional' = 'true','orc.compress'='ZLIB','orc.create.index'='true');

TRUNCATE TABLE assessdb.datetime003;

INSERT INTO TABLE assessdb.datetime003
SELECT
  CAST(id as INT), SUBSTRING(datetimes,2,LENGTH(datetimes)-2),
  SUBSTRING(monthname,2,LENGTH(monthname)-2), CAST(yearnumber as INT),
  CAST(monthnumber as INT),  CAST(daynumber as INT), CAST(hournumber as INT),
  CAST(minutenumber as INT), SUBSTRING(ampm,2,LENGTH(ampm)-2)
FROM assessdb.datetime002;

CREATE TABLE IF NOT EXISTS assessdb.dates (
  id          int, datetimes     string, monthname string,
  yearnumber int, monthnumber   int,    daynumber int,
  hournumber int, minutenumber  int,    ampm       string
)
CLUSTERED BY (id) INTO 1 BUCKETS
STORED AS orc
TBLPROPERTIES('transactional' = 'true','orc.compress'='ZLIB','orc.create.index'='true');

TRUNCATE TABLE assessdb.dates;

INSERT INTO TABLE assessdb.dates
SELECT
  id, datetimes, monthname, yearnumber, monthnumber, daynumber,
  hournumber, minutenumber, ampm
FROM assessdb.datetime003;
```

That was easy as you have mastered the basic rules. You are not bound by the size of the data set.

## Cleanup Assess Database

The next step is tidying up the assess layer.

```
DROP TABLE assessdb.datetime001;
DROP TABLE assessdb.datetime002;
DROP TABLE assessdb.datetime003;
```

See example script Assess003.txt for the Hive code.

## Create the assess Address Tables

Next you will master a "wider" data set using the address data.

---

You have the skills, and you just need to apply the rules you have mastered.

---

```
CREATE TABLE IF NOT EXISTS assessdb.address001 (
  id STRING, postcode STRING, latitude STRING, longitude STRING,
  easting STRING,northing STRING, gridref STRING, district STRING,
  ward STRING, districtcode STRING, wardcode STRING, country STRING,
  countycode STRING, constituency STRING, typearea STRING
)
CLUSTERED BY (id) INTO 1 BUCKETS
STORED AS orc
TBLPROPERTIES('transactional' = 'true','orc.compress'='ZLIB','orc.create.index'='true');

TRUNCATE TABLE assessdb.address001;

INSERT INTO TABLE assessdb.address001
SELECT
  id, postcode, latitude, longitude, easting, northing, gridref, district,
  ward, districtcode, wardcode, country, countycode, constituency, typearea
FROM retrievedb.rawaddress
WHERE id <> '"id"';

CREATE TABLE IF NOT EXISTS assessdb.address002 (
  id STRING, postcode STRING, latitude STRING, longitude STRING,
  easting STRING, northing STRING, gridref STRING, district STRING,
  ward STRING, districtcode STRING, wardcode STRING, country STRING,
  countycode STRING, constituency STRING, typearea STRING
)
CLUSTERED BY (id) INTO 1 BUCKETS
STORED AS orc
TBLPROPERTIES('transactional' = 'true','orc.compress'='ZLIB','orc.create.index'='true');

TRUNCATE TABLE assessdb.address002;

INSERT INTO TABLE assessdb.address002
SELECT
  id, rtrim(ltrim(postcode)), rtrim(ltrim(latitude)), rtrim(ltrim(longitude)),
  rtrim(ltrim(easting)), rtrim(ltrim(northing)), rtrim(ltrim(gridref)),
  rtrim(ltrim(district)), rtrim(ltrim(ward)), rtrim(ltrim(districtcode)),
  rtrim(ltrim(wardcode)), rtrim(ltrim(country)), rtrim(ltrim(countycode)),
  rtrim(ltrim(constituency)), rtrim(ltrim(typearea))
FROM assessdb.address001;

CREATE TABLE IF NOT EXISTS assessdb.address003 (
  id INT, postcode STRING, latitude DECIMAL(18, 9), longitude DECIMAL(18, 9),
  easting INT, northing INT, gridref STRING, district STRING, ward STRING,
```

```
    districtcode STRING, wardcode STRING, country STRING, countycode STRING,
    constituency STRING, typearea STRING
)
CLUSTERED BY (id) INTO 1 BUCKETS
STORED AS orc
TBLPROPERTIES('transactional' = 'true','orc.compress'='ZLIB','orc.create.index'='true');


TRUNCATE TABLE assessdb.address003;

INSERT INTO TABLE assessdb.address003
SELECT
    CAST(id as INT), SUBSTRING(postcode,2,LENGTH(postcode)-2),
    CAST(latitude as DECIMAL(18, 9)), CAST(longitude as DECIMAL(18, 9)),
    CAST(easting as INT), CAST(northing as INT),
    SUBSTRING(gridref,2,LENGTH(gridref)-2),
    SUBSTRING(district,2,LENGTH(district)-2),
    SUBSTRING(ward,2,LENGTH(ward)-2),
    SUBSTRING(districtcode,2,LENGTH(districtcode)-2),
    SUBSTRING(wardcode,2,LENGTH(wardcode)-2),
    SUBSTRING(country,2,LENGTH(country)-2),
    SUBSTRING(countycode,2,LENGTH(countycode)-2),
    SUBSTRING(constituency,2,LENGTH(constituency)-2),
    SUBSTRING(typearea,2,LENGTH(typearea)-2)
FROM assessdb.address002;

CREATE TABLE IF NOT EXISTS assessdb.postaddress (
    id INT, postcode STRING, latitude DECIMAL(18, 9),
    longitude DECIMAL(18, 9), easting INT, northing INT,
    gridref STRING, district STRING, ward STRING, districtcode STRING,
    wardcode STRING, country STRING, countycode STRING,
    constituency STRING, typearea STRING
)
CLUSTERED BY (id) INTO 1 BUCKETS
STORED AS orc
TBLPROPERTIES('transactional' = 'true','orc.compress'='ZLIB','orc.create.index'='true');


INSERT INTO TABLE assessdb.postaddress
SELECT
    id, postcode, latitude, longitude, easting, northing, gridref, district,
    ward, districtcode, wardcode, country, countycode, constituency, typearea
  FROM
    assessdb.address003;

CREATE TABLE IF NOT EXISTS assessdb.addresshistory001 (
    id STRING, pid STRING, aid STRING, did1 STRING, did2 STRING
)
CLUSTERED BY (id) INTO 1 BUCKETS
STORED AS orc
TBLPROPERTIES('transactional' = 'true','orc.compress'='ZLIB','orc.create.index'='true');
```

```
TRUNCATE TABLE assessdb.addresshistory001;

INSERT INTO TABLE assessdb.addresshistory001
SELECT
  id, pid, aid, did1, did2
FROM
  retrievedb.rawaddresshistory
WHERE id <> '"id"';

CREATE TABLE IF NOT EXISTS assessdb.addresshistory002 (
  id INT, pid INT, aid INT, did1 INT, did2 INT
)
CLUSTERED BY (id) INTO 1 BUCKETS
STORED AS orc
TBLPROPERTIES('transactional' = 'true','orc.compress'='ZLIB','orc.create.index'='true');

TRUNCATE TABLE assessdb.addresshistory002;

INSERT INTO TABLE assessdb.addresshistory002
SELECT
  CAST(id as INT), CAST(pid as INT), CAST(aid as INT),
  CAST(did1 as INT), CAST(did2 as INT)
FROM
  assessdb.addresshistory001;

CREATE TABLE IF NOT EXISTS assessdb.addresshistory (
  id INT, pid INT, aid INT, did1 INT, did2 INT
)
CLUSTERED BY (id) INTO 1 BUCKETS
STORED AS orc
TBLPROPERTIES('transactional' = 'true','orc.compress'='ZLIB','orc.create.index'='true');

TRUNCATE TABLE assessdb.addresshistory;

INSERT INTO TABLE assessdb.addresshistory
SELECT
  id, pid, aid, did1, did2
FROM
  assessdb.addresshistory002;
```

Once more, the number of fields has no impact on the rules. Just keep on applying them against the data sets.

## Clean Up the address Tables

```
DROP TABLE assessdb.address001;
DROP TABLE assessdb.address002;
DROP TABLE assessdb.address003;

DROP TABLE assessdb.addresshistory001;
DROP TABLE assessdb.addresshistory002;
```

## Evaluate the address Tables

```
SELECT
  addresshistory.id, addresshistory.pid, personfull.firstname,
  personfull.lastname, addresshistory.aid, postaddress.postcode,
  addresshistory.did1, dates1.datetimes as startdate,
  addresshistory.did2, dates2.datetimes as enddate
FROM
  assessdb.addresshistory
JOIN
  assessdb.personfull ON addresshistory.pid = personfull.persid
JOIN
  assessdb.postaddress ON addresshistory.aid = postaddress.id
JOIN
  assessdb.dates as dates1 ON addresshistory.did1 = dates1.id
JOIN
  assessdb.dates as dates2 ON addresshistory.did2 = dates2.id
LIMIT 20;
```

You can now see 20 records if you created the address data warehouse section. Let's load more data. You should have the process mastered.

---

See example script `Assess004.txt` for the Hive code.

---

## Create the assess account Tables

```
CREATE TABLE IF NOT EXISTS assessdb.account001 (
  id STRING, pid STRING, accountno STRING, balance STRING
)
CLUSTERED BY (id) INTO 1 BUCKETS
STORED AS orc
TBLPROPERTIES('transactional' = 'true','orc.compress'='ZLIB','orc.create.index'='true');

TRUNCATE TABLE assessdb.account001;

INSERT INTO TABLE assessdb.account001
SELECT
  id, pid, accountno, balance
FROM retrievedb.rawaccount
WHERE id <> '"id"';

CREATE TABLE IF NOT EXISTS assessdb.account002 (
  id STRING, pid STRING, accountno STRING, balance STRING
)
CLUSTERED BY (id) INTO 1 BUCKETS
STORED AS orc
TBLPROPERTIES('transactional' = 'true','orc.compress'='ZLIB','orc.create.index'='true');
```

```
TRUNCATE TABLE assessdb.account002;

INSERT INTO TABLE assessdb.account002
SELECT
  id, pid, rtrim(ltrim(accountno)), balance
FROM assessdb.account001;

CREATE TABLE IF NOT EXISTS assessdb.account003 (
  id INT, pid INT, accountid INT, accountno string, balance DECIMAL(18, 9)
)
CLUSTERED BY (id) INTO 1 BUCKETS
STORED AS orc
TBLPROPERTIES('transactional' = 'true','orc.compress'='ZLIB','orc.create.index'='true');
TRUNCATE TABLE assessdb.account003;

INSERT INTO TABLE assessdb.account003
SELECT
  CAST(id as INT), CAST(pid as INT), CAST(accountno as INT),
  CONCAT('AC',accountno), CAST(balance as DECIMAL(18, 9))
FROM assessdb.account002;

CREATE TABLE IF NOT EXISTS assessdb.account (
  id INT, pid INT, accountid INT, accountno STRING, balance DECIMAL(18, 9)
)
CLUSTERED BY (id) INTO 1 BUCKETS
STORED AS orc
TBLPROPERTIES('transactional' = 'true','orc.compress'='ZLIB','orc.create.index'='true');

TRUNCATE TABLE assessdb.account;

INSERT INTO TABLE assessdb.account
SELECT
  id, pid, accountid, accountno, balance
 FROM
  assessdb.account003;
```

## Clean Up the assess account Tables

```
DROP TABLE assessdb.account001;
DROP TABLE assessdb.account002;
DROP TABLE assessdb.account003;
```

You have now completed the assess layer for this book. Well done.

---

If you investigate the functions in Appendix B, you can master the functions to handle any corrections you need to make to the data during the assess layer's processing.

---

You can now proceed to the next layer.

# Process Database

The process database is structured as a data vault. Designed by Dan Linstedt, this database-modeling technique provides long-term chronological storage of data (see Figure 8-20).

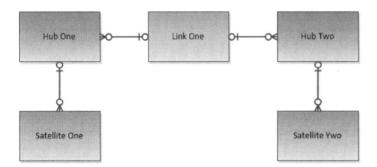

***Figure 8-20.*** *Basic data vault structure.*

The basic structure consists of three structures:

- *Hubs*—Comprise a list of unique business keys with little tendency to change.

- *Links*—Associations and transactions between business keys are recorded as links. These structures handle the relationships within the data set.

- *Satellites*—The hubs and links form the core structure of the data vault, but detail attributes are stored in isolated tables called satellites.

---

For more information, research the concepts of "data vaults".

---

Create a database called processdb to hold the process data structures:

```
CREATE DATABASE IF NOT EXISTS processdb;
```

The first table you create is personhub. The hub consists of:

- A hub key called id.

- A business key called keyid.

- Two natural keys called firstname and lastname.

```
USE processdb;

CREATE TABLE IF NOT EXISTS processdb.personhub (
  id        INT,
  keyid     STRING,
  firstname STRING,
  lastname  STRING
)
```

```
CLUSTERED BY (id) INTO 1 BUCKETS
STORED AS orc
TBLPROPERTIES('transactional' = 'true','orc.compress'='ZLIB','orc.create.index'='true');
```

You should use several keys to ensure you can handle any data restructuring in the future, if you have any issues with data integrity or have to rebuild the hub. In this example, you use Keyid and firstname plus lastname as two different keys for the same data set.

The second table you create is called personsexsatellite. The satellite consists of:

- A hub key called id.

- A business key called keyid from personhub.

- An attribute called sex.

- A timestamp called timestamp to record when data was loaded.

```
CREATE TABLE IF NOT EXISTS processdb.personsexsatellite (
  id         INT,
  keyid      STRING,
  sex        STRING,
  timestmp   BIGINT
)
CLUSTERED BY (id) INTO 1 BUCKETS
STORED AS orc
TBLPROPERTIES('transactional' = 'true','orc.compress'='ZLIB','orc.create.index'='true');
```

The third table you create is called person_person_link. It creates the business relationship of one person's relationship to another person.

The link consists of:

- A link key called id.

- A person hub key called personid1.

- A person hub key called personid2.

Here is the Hive code:

```
CREATE TABLE IF NOT EXISTS processdb.person_person_link(
  id INT,
  personid1 INT,
  personid2 INT
)
CLUSTERED BY (id, personid1, personid2) INTO 1 BUCKETS
STORED As orc
TBLPROPERTIES('transactional' = 'true','orc.compress'='ZLIB','orc.create.index'='true');
```

See example script Process001.txt for the Hive code. It holds the process-related data structures.

You can work through the code with ease now, as you have used the Hive code before and you are simply creating different data structures.

```
DROP DATABASE processdb CASCADE;

CREATE DATABASE IF NOT EXISTS processdb;
USE processdb;

CREATE TABLE IF NOT EXISTS processdb.personhub (
  id         INT,
  keyid      STRING,
  firstname  STRING,
  lastname   STRING
)
CLUSTERED BY (id) INTO 1 BUCKETS
STORED AS orc
TBLPROPERTIES('transactional' = 'true','orc.compress'='ZLIB','orc.create.index'='true');

CREATE TABLE IF NOT EXISTS processdb.personhub001 (
  firstname  STRING,
  lastname   STRING
)
CLUSTERED BY (firstname, lastname) INTO 1 BUCKETS
STORED AS orc
TBLPROPERTIES('transactional' = 'true','orc.compress'='ZLIB','orc.create.index'='true');

TRUNCATE TABLE processdb.personhub001;

INSERT INTO TABLE processdb.personhub001
SELECT DISTINCT
  firstname,
  lastname
FROM
  assessdb.personfull;

CREATE TABLE IF NOT EXISTS processdb.personhub002 (
  rid        BIGINT,
  tid        BIGINT,
  firstname  STRING,
  lastname   STRING
)
CLUSTERED BY (rid, tid) INTO 1 BUCKETS
STORED AS orc
TBLPROPERTIES('transactional' = 'true','orc.compress'='ZLIB','orc.create.index'='true');

TRUNCATE TABLE processdb.personhub002;

INSERT INTO TABLE processdb.personhub002
SELECT
  ROW_NUMBER() OVER (ORDER BY firstname, lastname),
  unix_timestamp(),
```

```
  firstname,
  lastname
FROM
  processdb.personhub001;

CREATE TABLE IF NOT EXISTS processdb.personhub003 (
  keyid        STRING,
  firstname    STRING,
  lastname     STRING
)
CLUSTERED BY (keyid) INTO 1 BUCKETS
STORED AS orc
TBLPROPERTIES('transactional' = 'true','orc.compress'='ZLIB','orc.create.index'='true');

TRUNCATE TABLE processdb.personhub003;

INSERT INTO TABLE processdb.personhub003
SELECT
  CONCAT(tid, '/', rid),
  firstname,
  lastname
FROM
  processdb.personhub002;

CREATE TABLE IF NOT EXISTS processdb.personhub004 (
  keyid        STRING,
  firstname    STRING,
  lastname     STRING,
  CDC          STRING
)
CLUSTERED BY (keyid) INTO 1 BUCKETS
STORED AS orc
TBLPROPERTIES('transactional' = 'true','orc.compress'='ZLIB','orc.create.index'='true');

TRUNCATE TABLE processdb.personhub004;

INSERT INTO TABLE processdb.personhub004
SELECT
  A.keyid,
  A.firstname,
  A.lastname,
  B.keyid
FROM
  processdb.personhub003 AS A
LEFT JOIN
  processdb.personhub AS B
ON
  A.firstname = B.firstname  AND A.lastname = B.lastname;

CREATE TABLE IF NOT EXISTS processdb.personhub005 (
  keyid        STRING,
```

```
   firstname    STRING,
   lastname     STRING
)
CLUSTERED BY (keyid) INTO 1 BUCKETS
STORED AS orc
TBLPROPERTIES('transactional' = 'true','orc.compress'='ZLIB','orc.create.index'='true');


TRUNCATE TABLE processdb.personhub005;


INSERT INTO TABLE processdb.personhub005
SELECT
  keyid,
  firstname,
  lastname
FROM
  processdb.personhub004
WHERE CDC IS NULL;


INSERT INTO TABLE processdb.personhub005
SELECT
  keyid,
  firstname,
  lastname
FROM
  processdb.personhub;


TRUNCATE TABLE processdb.personhub;


INSERT INTO TABLE processdb.personhub
SELECT
  ROW_NUMBER() OVER (ORDER BY keyid),
  keyid,
  firstname,
  lastname
FROM
  processdb.personhub005;


DROP TABLE processdb.personhub001;
DROP TABLE processdb.personhub002;
DROP TABLE processdb.personhub003;
DROP TABLE processdb.personhub004;


CREATE TABLE IF NOT EXISTS processdb.personsexsatellite001 (
  keyid        STRING,
  sex          STRING
)
CLUSTERED BY (keyid) INTO 1 BUCKETS
STORED AS orc
TBLPROPERTIES('transactional' = 'true','orc.compress'='ZLIB','orc.create.index'='true');
```

```
TRUNCATE TABLE processdb.personsexsatellite001;

INSERT INTO TABLE processdb.personsexsatellite001
SELECT DISTINCT
  A.keyid,
  B.sex
FROM
  processdb.personhub005 as A
JOIN
  assessdb.personfull AS B
ON
  A.firstname = B.firstname AND A.lastname = B.lastname;

CREATE TABLE IF NOT EXISTS processdb.personsexsatellite (
  id          INT,
  keyid       STRING,
  sex         STRING,
  timestmp    BIGINT
)
CLUSTERED BY (id) INTO 1 BUCKETS
STORED AS orc
TBLPROPERTIES('transactional' = 'true','orc.compress'='ZLIB','orc.create.index'='true');

TRUNCATE TABLE processdb.personsexsatellite;

INSERT INTO TABLE processdb.personsexsatellite
SELECT
  ROW_NUMBER() OVER (ORDER BY keyid),
  keyid,
  sex,
  unix_timestamp()
FROM
  processdb.personsexsatellite001;

DROP TABLE processdb.objecthub001;
DROP TABLE processdb.personsexsatellite001;
```

You are making good progress with your process layer. Well done!

---

See example script `Process002.txt` for the Hive code. It holds all the object-related data structures.

---

```
USE processdb;

CREATE TABLE IF NOT EXISTS processdb.objecthub (
  id          int,
  objecttype  string,
  objectname  string,
  objectid    int
)
```

```
CLUSTERED BY (id) INTO 1 BUCKETS
STORED AS orc
TBLPROPERTIES('transactional' = 'true','orc.compress'='ZLIB','orc.create.index'='true');

TRUNCATE TABLE processdb.objecthub;

CREATE TABLE IF NOT EXISTS processdb.objecthub001 (
  objecttype   string,
  objectname   string,
  objectid     int
)
CLUSTERED BY (objecttype, objectname,objectid) INTO 1 BUCKETS
STORED AS orc
TBLPROPERTIES('transactional' = 'true','orc.compress'='ZLIB','orc.create.index'='true');

TRUNCATE TABLE processdb.objecthub001;

INSERT INTO TABLE processdb.objecthub001
SELECT DISTINCT
  'intangible',
  'bankaccount',
  accountid
FROM
  assessdb.account;

TRUNCATE TABLE processdb.objecthub;

INSERT INTO TABLE processdb.objecthub
SELECT DISTINCT
  ROW_NUMBER() OVER (ORDER BY objecttype,objectname,objectid),
  objecttype,
  objectname,
  objectid
FROM
  processdb.objecthub001;

CREATE TABLE IF NOT EXISTS processdb.objectbankaccountsatellite0001 (
  accountid            int,
  transactionid        int,
  balance              DECIMAL(18, 9)
)
CLUSTERED BY (accountid,transactionid) INTO 1 BUCKETS
STORED AS orc
TBLPROPERTIES('transactional' = 'true','orc.compress'='ZLIB','orc.create.index'='true');

TRUNCATE TABLE processdb.objectbankaccountsatellite001;

INSERT INTO TABLE processdb.objectbankaccountsatellite0001
SELECT
  accountid,
  id as transactionid,
  balance
```

```
FROM
  assessdb.account;

CREATE TABLE IF NOT EXISTS processdb.objectbankaccountsatellite (
  id                int,
  accountid         int,
  transactionid     int,
  balance           DECIMAL(18, 9),
  timestmp          bigint
)
CLUSTERED BY (id) INTO 1 BUCKETS
STORED AS orc
TBLPROPERTIES('transactional' = 'true','orc.compress'='ZLIB','orc.create.index'='true');

TRUNCATE TABLE processdb.objectbankaccountsatellite;

INSERT INTO TABLE processdb.objectbankaccountsatellite
SELECT
  ROW_NUMBER() OVER (ORDER BY accountid,transactionid),
  accountid,
  transactionid,
  balance,
  unix_timestamp()
FROM
  processdb.objectbankaccountsatellite0001;

DROP TABLE processdb.objectbankaccountsatellite0001;
DROP TABLE processdb.objecthub001;
```

More progress ... Just keep on running the Hive code.

---

See example script Process003.txt for the Hive code. It holds all the location-related data structures.

---

```
USE processdb;

CREATE TABLE IF NOT EXISTS processdb.locationhub (
  id            INT,
  locationtype  STRING,
  locationname  STRING,
  locationid    INT
)
CLUSTERED BY (id) INTO 1 BUCKETS
STORED AS orc
TBLPROPERTIES('transactional' = 'true','orc.compress'='ZLIB','orc.create.index'='true');

TRUNCATE processdb.locationhub;

CREATE TABLE IF NOT EXISTS processdb.locationhub001 (
  locationtype  STRING,
  locationname  STRING,
```

```
    locationid    INT
)
CLUSTERED BY (locationtype, locationname,locationid) INTO 1 BUCKETS
STORED AS orc
TBLPROPERTIES('transactional' = 'true','orc.compress'='ZLIB','orc.create.index'='true');


TRUNCATE TABLE processdb.locationhub001;


INSERT INTO TABLE processdb.locationhub001
SELECT DISTINCT
  'intangible',
  'geospace',
  id as locationid
FROM
  assessdb.postaddress;


TRUNCATE TABLE processdb.locationhub;


INSERT INTO TABLE processdb.locationhub
SELECT DISTINCT
  ROW_NUMBER() OVER (ORDER BY locationtype,locationname,locationid),
  locationtype,
  locationname,
  locationid
FROM
  processdb.locationhub001;

CREATE TABLE IF NOT EXISTS processdb.locationgeospacesatellite0001 (
    locationid    INT,              postcode      STRING,
    latitude      DECIMAL(18, 9),   longitude     DECIMAL(18, 9),
    easting       INT,              northing      INT,
    gridref       STRING,           district      STRING,
    ward          STRING,           districtcode  STRING,
    wardcode      STRING,           country       STRING,
    countycode    STRING,           constituency  STRING,
    typearea      STRING
)
CLUSTERED BY (locationid) INTO 1 BUCKETS
STORED AS orc
TBLPROPERTIES('transactional' = 'true','orc.compress'='ZLIB','orc.create.index'='true');

TRUNCATE TABLE processdb.locationgeospacesatellite0001;

INSERT INTO TABLE processdb.locationgeospacesatellite0001
SELECT
  id as locationid, postcode, latitude, longitude, easting, northing,gridref,
  district, ward, districtcode, wardcode, country, countycode,
  constituency, typearea
FROM
  assessdb.postaddress;
  CREATE TABLE IF NOT EXISTS processdb.locationgeospace1satellite (
  id            INT,
```

```
  locationid    INT,
  postcode      STRING,
  timestmp      BIGINT
)
CLUSTERED BY (id) INTO 1 BUCKETS
STORED AS orc
TBLPROPERTIES('transactional' = 'true','orc.compress'='ZLIB','orc.create.index'='true');

TRUNCATE TABLE processdb.locationgeospace1satellite;

INSERT INTO TABLE processdb.locationgeospace1satellite
SELECT
  ROW_NUMBER() OVER (ORDER BY locationid),
  locationid,
  postcode,
  unix_timestamp()
FROM
  processdb.locationgeospacesatellite0001
ORDER BY locationid;

CREATE TABLE IF NOT EXISTS processdb.locationgeospace2satellite (
  id            INT,
  locationid    INT,
  latitude      DECIMAL(18, 9),
  longitude     DECIMAL(18, 9),
  timestmp      BIGINT
)
CLUSTERED BY (id, locationid) INTO 1 BUCKETS
STORED AS orc
TBLPROPERTIES('transactional' = 'true','orc.compress'='ZLIB','orc.create.index'='true');

TRUNCATE TABLE processdb.locationgeospace2satellite;

INSERT INTO TABLE processdb.locationgeospace2satellite
SELECT
  ROW_NUMBER() OVER (ORDER BY locationid),
  locationid,
  latitude,
  longitude,
  unix_timestamp()
FROM
  processdb.locationgeospacesatellite0001;

CREATE TABLE IF NOT EXISTS processdb.locationgeospace3satellite (
  id            INT,
  locationid    INT,
  easting       INT,
  northing      INT,
  timestmp      BIGINT
)
CLUSTERED BY (id, locationid) INTO 1 BUCKETS
STORED AS orc
```

```
TBLPROPERTIES('transactional' = 'true','orc.compress'='ZLIB','orc.create.index'='true');

TRUNCATE TABLE processdb.locationgeospace3satellite;

INSERT INTO TABLE processdb.locationgeospace3satellite
SELECT
  ROW_NUMBER() OVER (ORDER BY locationid),
  locationid,
  easting,
  northing,
  unix_timestamp()
FROM
  processdb.locationgeospacesatellite0001;

CREATE TABLE IF NOT EXISTS processdb.locationgeospace4satellite (
  id            INT,
  locationid    INT,
  postcode      STRING,
  latitude      DECIMAL(18, 9),
  longitude     DECIMAL(18, 9),
  easting       INT,
  northing      INT,
  gridref       STRING,
  district      STRING,
  ward          STRING,
  districtcode  STRING,
  wardcode      STRING,
  country       STRING,
  countycode    STRING,
  constituency  STRING,
  typearea      STRING,
  timestmp      BIGINT
)
CLUSTERED BY (id, locationid) INTO 1 BUCKETS
STORED AS orc
TBLPROPERTIES('transactional' = 'true','orc.compress'='ZLIB','orc.create.index'='true');

TRUNCATE TABLE processdb.locationgeospace4satellite;

INSERT INTO TABLE processdb.locationgeospace4satellite
SELECT
  ROW_NUMBER() OVER (ORDER BY locationid),
  locationid,
  postcode,
  latitude,
  longitude,
  easting,
  northing,
  gridref,
  district,
  ward,
```

```
  districtcode,
  wardcode,
  country,
  countycode,
  constituency,
  typearea,
  unix_timestamp()
FROM
  processdb.locationgeospacesatellite0001;

DROP TABLE processdb.locationgeospacesatellite0001;
DROP TABLE processdb.locationhub001;
```

We are nearly finished ... some more structures are required.

---

See example script `Process004.txt` for the Hive code. It holds all the event-related data structures.

---

```
USE processdb;

CREATE TABLE IF NOT EXISTS processdb.eventhub (
  id         int,
  eventtype  string,
  eventname  string,
  eventid    int
)
CLUSTERED BY (id) INTO 1 BUCKETS
STORED AS orc
TBLPROPERTIES('transactional' = 'true','orc.compress'='ZLIB','orc.create.index'='true');

TRUNCATE processdb.eventhub;

CREATE TABLE IF NOT EXISTS processdb.eventhub001 (
  eventtype  string,
  eventname  string,
  eventid    int
)
CLUSTERED BY (eventtype, eventname,eventid) INTO 1 BUCKETS
STORED AS orc
TBLPROPERTIES('transactional' = 'true','orc.compress'='ZLIB','orc.create.index'='true');

TRUNCATE TABLE processdb.eventhub001;

INSERT INTO TABLE processdb.eventhub001
SELECT DISTINCT
  'intangible',
  'banktransaction',
  id as eventid
FROM
  assessdb.account;
```

```
TRUNCATE TABLE processdb.eventhub;

INSERT INTO TABLE processdb.eventhub
SELECT DISTINCT
  ROW_NUMBER() OVER (ORDER BY eventtype,eventname,eventid),
  eventtype,
  eventname,
  eventid
FROM
  processdb.eventhub001;

CREATE TABLE IF NOT EXISTS processdb.eventbanktransactionsatellite0001 (
  accountid           int,
  transactionid       int,
  balance             DECIMAL(18, 9)
)
CLUSTERED BY (accountid,transactionid) INTO 1 BUCKETS
STORED AS orc
TBLPROPERTIES('transactional' = 'true','orc.compress'='ZLIB','orc.create.index'='true');

TRUNCATE TABLE processdb.eventbanktransactionsatellite001;

INSERT INTO TABLE processdb.eventbanktransactionsatellite0001
SELECT
  accountid,
  id as transactionid,
  balance
FROM
  assessdb.account;
CREATE TABLE IF NOT EXISTS processdb.eventbanktransactionsatellite (
  id                  int,
  accountid           int,
  transactionid       int,
  balance             DECIMAL(18, 9),
  timestmp            bigint
)
CLUSTERED BY (id) INTO 1 BUCKETS
STORED AS orc
TBLPROPERTIES('transactional' = 'true','orc.compress'='ZLIB','orc.create.index'='true');

TRUNCATE TABLE processdb.eventbanktransactionsatellite;

INSERT INTO TABLE processdb.eventbanktransactionsatellite
SELECT
  ROW_NUMBER() OVER (ORDER BY accountid,transactionid),
  accountid,
  transactionid,
  balance,
  unix_timestamp()
FROM
  processdb.eventbanktransactionsatellite0001;
```

```
DROP TABLE processdb.eventbanktransactionsatellite0001;
DROP TABLE processdb.eventhub001;
SHOW TABLES;
```

See example script Process005.txt for the Hive code. It holds all the time-related data structures.

```
USE processdb;

CREATE TABLE IF NOT EXISTS processdb.timehub (
  id             INT,
  timeid     INT
)
CLUSTERED BY (id) INTO 1 BUCKETS
STORED AS orc
TBLPROPERTIES('transactional' = 'true','orc.compress'='ZLIB','orc.create.index'='true');

TRUNCATE TABLE processdb.timehub;
CREATE TABLE IF NOT EXISTS processdb.timehub001 (
  timeid     INT
)
CLUSTERED BY (timeid) INTO 1 BUCKETS
STORED AS orc
TBLPROPERTIES('transactional' = 'true','orc.compress'='ZLIB','orc.create.index'='true');

TRUNCATE TABLE processdb.timehub001;

INSERT INTO TABLE processdb.timehub001
SELECT DISTINCT
  id as timeid
FROM
  assessdb.dates
WHERE yearnumber = 2015;

INSERT INTO TABLE processdb.timehub001
SELECT DISTINCT
  id as timeid
FROM
  assessdb.dates
WHERE yearnumber = 2016;

TRUNCATE TABLE processdb.timehub;

INSERT INTO TABLE processdb.timehub
SELECT DISTINCT
  ROW_NUMBER() OVER (ORDER BY timeid),
  timeid
FROM
  processdb.timehub001;
```

```
CREATE TABLE IF NOT EXISTS processdb.timesatellite0001 (
  timeid        INT,
  datetimes     string
)
CLUSTERED BY (timeid) INTO 1 BUCKETS
STORED AS orc
TBLPROPERTIES('transactional' = 'true','orc.compress'='ZLIB','orc.create.index'='true');

TRUNCATE TABLE processdb.timesatellite0001;

INSERT INTO TABLE processdb.timesatellite0001
SELECT
  id as timeid,
  datetimes
FROM
  assessdb.dates
WHERE yearnumber = 2015;

INSERT INTO TABLE processdb.timesatellite0001
SELECT
  id as timeid,
  datetimes
FROM
  assessdb.dates
WHERE yearnumber = 2016;

CREATE TABLE IF NOT EXISTS processdb.time1satellite (
  id            INT,
  timeid        INT,
  datetimes     STRING,
  timestmp      BIGINT
)
CLUSTERED BY (id) INTO 1 BUCKETS
STORED AS orc
TBLPROPERTIES('transactional' = 'true','orc.compress'='ZLIB','orc.create.index'='true');

TRUNCATE TABLE processdb.time1satellite;

INSERT INTO TABLE processdb.time1satellite
SELECT
  ROW_NUMBER() OVER (ORDER BY timeid),
  timeid,
  datetimes,
  unix_timestamp()
FROM
  processdb.timesatellite0001
ORDER BY timeid;

DROP TABLE processdb.timesatellite0001;
DROP TABLE processdb.timehub001;
```

You have now created all the hubs and satellites. Now you will add all the link tables. This section is extensive, but the reward is close. You will soon have a fully working data vault.

---

See example script `Process006.txt` for the Hive code. It holds all the links between the person, object, location, event, and time data structures.

---

```
USE processdb;

CREATE TABLE IF NOT EXISTS processdb.person_person_link(
  id INT,
  personid1 INT,
  personid2 INT
)
CLUSTERED BY (id, personid1, personid2) INTO 1 BUCKETS
STORED As orc
TBLPROPERTIES('transactional' = 'true','orc.compress'='ZLIB','orc.create.index'='true');

TRUNCATE TABLE processdb.person_person_link;

CREATE TABLE IF NOT EXISTS processdb.person_person_link002(
  personid1 INT,
  personid2 INT
)
CLUSTERED BY (personid1, personid2) INTO 1 BUCKETS
STORED As orc
TBLPROPERTIES('transactional' = 'true','orc.compress'='ZLIB','orc.create.index'='true');

TRUNCATE TABLE processdb.person_person_link002;

CREATE TABLE IF NOT EXISTS processdb.personlink001(
  personid INT
)
CLUSTERED BY (personid) INTO 1 BUCKETS
STORED As orc
TBLPROPERTIES('transactional' = 'true','orc.compress'='ZLIB','orc.create.index'='true');

INSERT INTO TABLE processdb.personlink001
SELECT
  personhub.id as personid
FROM
  processdb.personhub
LIMIT 10;

CREATE TABLE IF NOT EXISTS processdb.object_object_link(
  id INT,
  objectid1 INT,
  objectid2 INT
)
CLUSTERED BY (id, objectid1, objectid2) INTO 1 BUCKETS
STORED As orc
TBLPROPERTIES('transactional' = 'true','orc.compress'='ZLIB','orc.create.index'='true');
```

```
CREATE TABLE IF NOT EXISTS processdb.object_object_link002(
  objectid1 INT,
  objectid2 INT
)
CLUSTERED BY (objectid1, objectid2) INTO 1 BUCKETS
STORED As orc
TBLPROPERTIES('transactional' = 'true','orc.compress'='ZLIB','orc.create.index'='true');

CREATE TABLE IF NOT EXISTS processdb.objectlink001(
  objectid INT
)
CLUSTERED BY (objectid) INTO 1 BUCKETS
STORED As orc
TBLPROPERTIES('transactional' = 'true','orc.compress'='ZLIB','orc.create.index'='true');

TRUNCATE TABLE processdb.objectlink001;

INSERT INTO TABLE processdb.objectlink001
SELECT
  objecthub.id as objectid
FROM
  processdb.objecthub
LIMIT 10;

CREATE TABLE IF NOT EXISTS processdb.location_location_link(
  id INT,
  locationid1 INT,
  locationid2 INT
)
CLUSTERED BY (id, locationid1, locationid2) INTO 1 BUCKETS
STORED As orc
TBLPROPERTIES('transactional' = 'true','orc.compress'='ZLIB','orc.create.index'='true');

TRUNCATE TABLE processdb.location_location_link;

CREATE TABLE IF NOT EXISTS processdb.location_location_link002(
  locationid1 INT,
  locationid2 INT
)
CLUSTERED BY (locationid1, locationid2) INTO 1 BUCKETS
STORED As orc
TBLPROPERTIES('transactional' = 'true','orc.compress'='ZLIB','orc.create.index'='true');

CREATE TABLE IF NOT EXISTS processdb.locationlink001(
  locationid INT
)
CLUSTERED BY (locationid) INTO 1 BUCKETS
STORED As orc
TBLPROPERTIES('transactional' = 'true','orc.compress'='ZLIB','orc.create.index'='true');

INSERT INTO TABLE processdb.locationlink001
```

```
SELECT
  locationhub.id as locationid
FROM
  processdb.locationhub
LIMIT 10;

CREATE TABLE IF NOT EXISTS processdb.event_event_link(
  id INT,
  eventid1 INT,
  eventid2 INT
)
CLUSTERED BY (id, eventid1, eventid2) INTO 1 BUCKETS
STORED As orc
TBLPROPERTIES('transactional' = 'true','orc.compress'='ZLIB','orc.create.index'='true');

CREATE TABLE IF NOT EXISTS processdb.event_event_link002(
  eventid1 INT,
  eventid2 INT
)
CLUSTERED BY (eventid1, eventid2) INTO 1 BUCKETS
STORED As orc
TBLPROPERTIES('transactional' = 'true','orc.compress'='ZLIB','orc.create.index'='true');

CREATE TABLE IF NOT EXISTS processdb.eventlink001(
  eventid INT
)
CLUSTERED BY (eventid) INTO 1 BUCKETS
STORED As orc
TBLPROPERTIES('transactional' = 'true','orc.compress'='ZLIB','orc.create.index'='true');
INSERT INTO TABLE processdb.eventlink001
SELECT
  eventhub.id as eventid
FROM
  processdb.eventhub
LIMIT 10;

CREATE TABLE IF NOT EXISTS processdb.time_time_link(
  id INT,
  timeid1 INT,
  timeid2 INT
)
CLUSTERED BY (id, timeid1, timeid2) INTO 1 BUCKETS
STORED As orc
TBLPROPERTIES('transactional' = 'true','orc.compress'='ZLIB','orc.create.index'='true');

CREATE TABLE IF NOT EXISTS processdb.time_time_link002(
  timeid1 INT,
  timeid2 INT
)
CLUSTERED BY (timeid1, timeid2) INTO 1 BUCKETS
STORED As orc
```

```
TBLPROPERTIES('transactional' = 'true','orc.compress'='ZLIB','orc.create.index'='true');

CREATE TABLE IF NOT EXISTS processdb.timelink001(
  timeid INT
)
CLUSTERED BY (timeid) INTO 1 BUCKETS
STORED As orc
TBLPROPERTIES('transactional' = 'true','orc.compress'='ZLIB','orc.create.index'='true');

INSERT INTO TABLE processdb.timelink001
SELECT
  timehub.id as timeid
FROM
  processdb.timehub
LIMIT 10;

CREATE TABLE IF NOT EXISTS processdb.person_object_link002(
  personid INT,
  objectid INT
)
CLUSTERED BY (personid, objectid) INTO 1 BUCKETS
STORED As orc
TBLPROPERTIES('transactional' = 'true','orc.compress'='ZLIB','orc.create.index'='true');

INSERT INTO TABLE processdb.person_object_link002
SELECT DISTINCT
  personlink001.id as personid,
  objectlink001.id as objectid
FROM
  processdb.personlink001
GROSS JOIN
  processdb.objectlink001
LIMIT 20;

INSERT INTO TABLE processdb.person_object_link002
SELECT personhub.id, objecthub.objectid
FROM assessdb.account
JOIN
processdb.personhub
ON account.pid = personhub.id
JOIN
processdb.objecthub
ON account.accountid = objecthub.objectid
LIMIT 100;

CREATE TABLE IF NOT EXISTS processdb.person_object_link(
  id INT,
  personid INT,
  objectid INT
)
CLUSTERED BY (id, personid, objectid) INTO 1 BUCKETS
```

```
STORED As orc
TBLPROPERTIES('transactional' = 'true','orc.compress'='ZLIB','orc.create.index'='true');

INSERT INTO TABLE processdb.person_object_link
SELECT DISTINCT
  ROW_NUMBER() OVER (ORDER BY personid, objectid),
  personid,
  objectid
FROM
  processdb.person_object_link002;

CREATE TABLE IF NOT EXISTS processdb.person_location_link002(
  personid INT,
  locationid INT
)
CLUSTERED BY (personid, locationid) INTO 1 BUCKETS
STORED As orc
TBLPROPERTIES('transactional' = 'true','orc.compress'='ZLIB','orc.create.index'='true');

INSERT INTO TABLE processdb.person_location_link002
SELECT DISTINCT
  personlink001.id as personid,
  locationlink001.id as locationid
FROM
  processdb.personlink001
GROSS JOIN
  processdb.locationlink001
LIMIT 20;

CREATE TABLE IF NOT EXISTS processdb.person_location_link(
  id INT,
  personid INT,
  locationid INT
)
CLUSTERED BY (id, personid, locationid) INTO 1 BUCKETS
STORED As orc
TBLPROPERTIES('transactional' = 'true','orc.compress'='ZLIB','orc.create.index'='true');

INSERT INTO TABLE processdb.person_location_link
SELECT DISTINCT
  ROW_NUMBER() OVER (ORDER BY personid, locationid),
  personid,
  locationid
FROM
  processdb.person_location_link002;

CREATE TABLE IF NOT EXISTS processdb.person_event_link002(
  personid INT,
  eventid INT
)
CLUSTERED BY (personid, eventid) INTO 1 BUCKETS
```

```
STORED As orc
TBLPROPERTIES('transactional' = 'true','orc.compress'='ZLIB','orc.create.index'='true');

INSERT INTO TABLE processdb.person_event_link002
SELECT DISTINCT
  personlink001.id as personid,
  eventlink001.id as eventid
FROM
  processdb.personlink001
GROSS JOIN
  processdb.eventlink001
LIMIT 20;

CREATE TABLE IF NOT EXISTS processdb.person_event_link(
  id INT,
  personid INT,
  eventid INT
)
CLUSTERED BY (id, personid, eventid) INTO 1 BUCKETS
STORED As orc
TBLPROPERTIES('transactional' = 'true','orc.compress'='ZLIB','orc.create.index'='true');

INSERT INTO TABLE processdb.person_event_link
SELECT DISTINCT
  ROW_NUMBER() OVER (ORDER BY personid, eventid),
  personid,
  eventid
FROM
  processdb.person_event_link002;

CREATE TABLE IF NOT EXISTS processdb.person_time_link002(
  personid INT,
  timeid INT
)
CLUSTERED BY (personid, timeid) INTO 1 BUCKETS
STORED As orc
TBLPROPERTIES('transactional' = 'true','orc.compress'='ZLIB','orc.create.index'='true');

INSERT INTO TABLE processdb.person_time_link002
SELECT DISTINCT
  personlink001.id as personid,
  timelink001.id as timeid
FROM
  processdb.personlink001
GROSS JOIN
  processdb.timelink001
LIMIT 20;

CREATE TABLE IF NOT EXISTS processdb.person_time_link(
  id INT,
  personid INT,
  timeid INT
```

```
)
CLUSTERED BY (id, personid, timeid) INTO 1 BUCKETS
STORED As orc
TBLPROPERTIES('transactional' = 'true','orc.compress'='ZLIB','orc.create.index'='true');

INSERT INTO TABLE processdb.person_time_link
SELECT DISTINCT
  ROW_NUMBER() OVER (ORDER BY personid, timeid),
  personid,
  timeid
FROM
  processdb.person_time_link002;

CREATE TABLE IF NOT EXISTS processdb.object_location_link002(
  objectid INT,
  locationid INT
)
CLUSTERED BY (objectid, locationid) INTO 1 BUCKETS
STORED As orc
TBLPROPERTIES('transactional' = 'true','orc.compress'='ZLIB','orc.create.index'='true');

INSERT INTO TABLE processdb.object_location_link002
SELECT DISTINCT
  objectlink001.id as objectid,
  locationlink001.id as locationid
FROM
  processdb.objectlink001
GROSS JOIN
  processdb.locationlink001
LIMIT 20;

CREATE TABLE IF NOT EXISTS processdb.object_location_link(
  id INT,
  objectid INT,
  locationid INT
)
CLUSTERED BY (id, objectid, locationid) INTO 1 BUCKETS
STORED As orc
TBLPROPERTIES('transactional' = 'true','orc.compress'='ZLIB','orc.create.index'='true');

INSERT INTO TABLE processdb.object_location_link
SELECT DISTINCT
  ROW_NUMBER() OVER (ORDER BY objectid, locationid),
  objectid,
  locationid
FROM
  processdb.object_location_link002;

CREATE TABLE IF NOT EXISTS processdb.object_event_link002(
  objectid INT,
  eventid INT
)
```

```
CLUSTERED BY (objectid, eventid) INTO 1 BUCKETS
STORED As orc
TBLPROPERTIES('transactional' = 'true','orc.compress'='ZLIB','orc.create.index'='true');

INSERT INTO TABLE processdb.object_event_link002
SELECT DISTINCT
  objectlink001.id as objectid,
  eventlink001.id as eventid
FROM
  processdb.objectlink001
GROSS JOIN
  processdb.eventlink001
LIMIT 20;

CREATE TABLE IF NOT EXISTS processdb.object_event_link(
  id INT,
  objectid INT,
  eventid INT
)
CLUSTERED BY (id, objectid, eventid) INTO 1 BUCKETS
STORED As orc
TBLPROPERTIES('transactional' = 'true','orc.compress'='ZLIB','orc.create.index'='true');

INSERT INTO TABLE processdb.object_event_link
SELECT DISTINCT
  ROW_NUMBER() OVER (ORDER BY objectid, eventid),
  objectid,
  eventid
FROM
  processdb.object_event_link002;

CREATE TABLE IF NOT EXISTS processdb.object_time_link002(
  objectid INT,
  timeid INT
)
CLUSTERED BY (objectid, timeid) INTO 1 BUCKETS
STORED As orc
TBLPROPERTIES('transactional' = 'true','orc.compress'='ZLIB','orc.create.index'='true');

INSERT INTO TABLE processdb.object_time_link002
SELECT DISTINCT
  objectlink001.id as objectid,
  timelink001.id as timeid
FROM
  processdb.objectlink001
GROSS JOIN
  processdb.timelink001
LIMIT 20;

CREATE TABLE IF NOT EXISTS processdb.object_time_link(
  id INT,
  objectid INT,
```

```
  timeid INT
)
CLUSTERED BY (id, objectid, timeid) INTO 1 BUCKETS
STORED As orc
TBLPROPERTIES('transactional' = 'true','orc.compress'='ZLIB','orc.create.index'='true');

INSERT INTO TABLE processdb.object_time_link
SELECT DISTINCT
  ROW_NUMBER() OVER (ORDER BY objectid, timeid),
  objectid,
  timeid
FROM
  processdb.object_time_link002;

CREATE TABLE IF NOT EXISTS processdb.location_event_link002(
  locationid INT,
  eventid INT
)
CLUSTERED BY (locationid, eventid) INTO 1 BUCKETS
STORED As orc
TBLPROPERTIES('transactional' = 'true','orc.compress'='ZLIB','orc.create.index'='true');

INSERT INTO TABLE processdb.location_event_link002
SELECT DISTINCT
  locationlink001.id as locationid,
  eventlink001.id as eventid
FROM
  processdb.locationlink001
GROSS JOIN
  processdb.eventlink001
LIMIT 20;

CREATE TABLE IF NOT EXISTS processdb.location_event_link(
  id INT,
  locationid INT,
  eventid INT
)
CLUSTERED BY (id, locationid, eventid) INTO 1 BUCKETS
STORED As orc
TBLPROPERTIES('transactional' = 'true','orc.compress'='ZLIB','orc.create.index'='true');

INSERT INTO TABLE processdb.location_event_link
SELECT DISTINCT
  ROW_NUMBER() OVER (ORDER BY locationid, eventid),
  locationid,
  eventid
FROM
  processdb.location_event_link002;

CREATE TABLE IF NOT EXISTS processdb.location_time_link002(
  locationid INT,
  timeid INT
```

```
)
CLUSTERED BY (locationid, timeid) INTO 1 BUCKETS
STORED As orc
TBLPROPERTIES('transactional' = 'true','orc.compress'='ZLIB','orc.create.index'='true');

INSERT INTO TABLE processdb.location_time_link002
SELECT DISTINCT
  locationlink001.id as locationid,
  timelink001.id as timeid
FROM
  processdb.locationlink001
GROSS JOIN
  processdb.timelink001
LIMIT 20;

CREATE TABLE IF NOT EXISTS processdb.location_time_link(
  id INT,
  locationid INT,
  timeid INT
)
CLUSTERED BY (id, locationid, timeid) INTO 1 BUCKETS
STORED As orc
TBLPROPERTIES('transactional' = 'true','orc.compress'='ZLIB','orc.create.index'='true');

INSERT INTO TABLE processdb.location_time_link
SELECT DISTINCT
  ROW_NUMBER() OVER (ORDER BY locationid, timeid),
  locationid,
  timeid
FROM
  processdb.location_time_link002;

CREATE TABLE IF NOT EXISTS processdb.event_time_link002(
  eventid INT,
  timeid INT
)
CLUSTERED BY (eventid, timeid) INTO 1 BUCKETS
STORED As orc
TBLPROPERTIES('transactional' = 'true','orc.compress'='ZLIB','orc.create.index'='true');

INSERT INTO TABLE processdb.event_time_link002
SELECT DISTINCT
  eventlink001.id as eventid,
  timelink001.id as timeid
FROM
  processdb.eventlink001
GROSS JOIN
  processdb.timelink001
LIMIT 20;

CREATE TABLE IF NOT EXISTS processdb.event_time_link(
```

```
  id INT,
  eventid INT,
  timeid INT
)
CLUSTERED BY (id, eventid, timeid) INTO 1 BUCKETS
STORED As orc
TBLPROPERTIES('transactional' = 'true','orc.compress'='ZLIB','orc.create.index'='true');

INSERT INTO TABLE processdb.event_time_link
SELECT DISTINCT
  ROW_NUMBER() OVER (ORDER BY eventid, timeid),
  eventid,
  timeid
FROM
  processdb.event_time_link002;
```

You now have a data vault. Let's just clean up the processdb database and we are done.

---

See example script Process007.txt for the Hive code. It cleans up the process database.

---

```
USE processdb;

DROP TABLE processdb.person_event_link002;
DROP TABLE processdb.person_location_link002;
DROP TABLE processdb.person_object_link002;
DROP TABLE processdb.person_person_link002;
DROP TABLE processdb.person_time_link002;
DROP TABLE processdb.personlink001;

DROP TABLE processdb.object_event_link002;
DROP TABLE processdb.object_location_link002;
DROP TABLE processdb.object_object_link002;
DROP TABLE processdb.object_time_link002;
DROP TABLE processdb.objectlink001;

DROP TABLE processdb.location_event_link002;
DROP TABLE processdb.location_location_link002;
DROP TABLE processdb.location_time_link002;
DROP TABLE processdb.locationlink001;

DROP TABLE processdb.event_event_link002;
DROP TABLE processdb.event_time_link002;
DROP TABLE processdb.eventlink001;

DROP TABLE processdb.time_time_link002;
DROP TABLE processdb.timelink001;
```

You have now completed the range of scripts against your Hive solution to create all the data structures for processdb.

Let's quickly verify which tables you have created. Execute this command:

```
SHOW TABLES;
```

Success! You have completed the process layer.

# Transform Database

The transform database holds a ROLAP (Relational Online Analytical Processing) model consisting of the physical deployment of the dimensions and facts as described by the sun models.

You create a database called transformdb to hold the transform data structures as recommended by your sun models.

```
CREATE DATABASE IF NOT EXISTS transformdb;
USE transformdb;
```

The first dimension you create is dimperson, which consists of:

- A dimension key called personkey.

- Two dimensional attributes called firstname and lastname.

```
CREATE TABLE IF NOT EXISTS transformdb.dimperson (
  personkey  BIGINT,
  firstname  STRING,
  lastname   STRING
)
CLUSTERED BY (firstname, lastname,personkey) INTO 1 BUCKETS
STORED AS orc
TBLPROPERTIES('transactional' = 'true','orc.compress'='ZLIB','orc.create.index'='true');
```

Let's load the sample data into the dimension for person.

```
INSERT INTO TABLE transformdb.dimperson
VALUES
(999997,'Ruff','Hond'),
(999998,'Robbie','Rot'),
(999999,'Helen','Kat');
```

---

▓ **Note**    We are simply inserting data because it speeds up the processing through this layer.

---

The second dimension you create is dimaccount, which consists of:

- A dimension key called accountkey.

- A dimensional attribute called accountnumber.

```
CREATE TABLE IF NOT EXISTS transformdb.dimaccount (
  accountkey     BIGINT,
  accountnumber  INT
)
```

```
CLUSTERED BY (accountnumber,accountkey) INTO 1 BUCKETS
STORED AS orc
TBLPROPERTIES('transactional' = 'true','orc.compress'='ZLIB','orc.create.index'='true');
```

Let's load some sample data into the dimension for the dimaccount.

```
INSERT INTO TABLE transformdb.dimaccount
VALUES
(88888887,208887),
(88888888,208888),
(88888889,208889);
```

The first fact you create is fctpersonaccount, which consists of the following:

- A fact key called personaccountkey.

- A fact key called personkey from dimension dimperson.

- A fact key called accountkey from dimension dimaccount.

- A measure called balance.

```
CREATE TABLE IF NOT EXISTS transformdb.fctpersonaccount (
  personaccountkey      BIGINT,
  personkey             BIGINT,
  accountkey            BIGINT,
  balance               DECIMAL(18, 9)
)
CLUSTERED BY (personkey,accountkey) INTO 1 BUCKETS
STORED AS orc
TBLPROPERTIES('transactional' = 'true','orc.compress'='ZLIB','orc.create.index'='true');
```

Let's load some sample data into the fact fctpersonaccount.
The next interim fact table you create is fctpersonaccount001:

```
CREATE TABLE IF NOT EXISTS transformdb.fctpersonaccount001 (
  personkey             BIGINT,
  accountkey            BIGINT,
  balance               DECIMAL(18, 9)
)
CLUSTERED BY (personkey,accountkey) INTO 1 BUCKETS
STORED AS orc
TBLPROPERTIES('transactional' = 'true','orc.compress'='ZLIB','orc.create.index'='true');

INSERT INTO TABLE transformdb.fctpersonaccount001
VALUES
(999997,88888887,10.60),
(999997,88888887,400.70),
(999997,88888887,-210.90),
(999998,88888888,1000.00),
(999998,88888888,1990.60),
(999998,88888888,900.70),
(999999,88888889,160.60),
```

```
(999999,88888889,180.70),
(999999,88888889,100.60),
(999999,88888889,120.90),
(999999,88888889,180.69),
(999999,88888889,130.30);
```

The next interim fact table you create is `fctpersonaccount002`:

```
CREATE TABLE IF NOT EXISTS transformdb.fctpersonaccount002 (
  personkey       BIGINT,
  accountkey      BIGINT,
  balance         DECIMAL(18, 9)
)
CLUSTERED BY (personkey,accountkey) INTO 1 BUCKETS
STORED AS orc
TBLPROPERTIES('transactional' = 'true','orc.compress'='ZLIB','orc.create.index'='true');
```

Let's load some active data into the fact `fctpersonaccount002`.

```
INSERT INTO TABLE transformdb.fctpersonaccount002
SELECT
CAST(personkey AS BIGINT),
CAST(accountkey AS BIGINT),
CAST(SUM(balance) AS DECIMAL(18, 9))
FROM transformdb.fctpersonaccount001
GROUP BY personkey, accountkey;
```

Let's load some active data into the fact `fctpersonaccount` by using the dimensions `dimperson` and `dimaccount` via fact `fctpersonaccount0002`.

```
INSERT INTO TABLE transformdb.fctpersonaccount
SELECT
ROW_NUMBER() OVER (ORDER BY personkey, accountkey),
CAST(personkey AS BIGINT),
CAST(accountkey AS BIGINT),
CAST(balance AS DECIMAL(18, 9))
FROM transformdb.fctpersonaccount002;
```

Clean up the `transformdb`:

```
DROP TABLE transformdb.fctpersonaccount001;
DROP TABLE transformdb.fctpersonaccount002;
```

You now have the basic building blocks for the transform ROLAP structures. Let's deploy your well mastered Hive skills against the transform requirements and build the complete transform database.

---

■ **Note**  See example script `Transform01.txt` for the Hive code. It creates and populates the dimension `dimperson`.

---

```
DROP DATABASE transformdb CASCADE;

CREATE DATABASE IF NOT EXISTS transformdb;
USE transformdb;

CREATE TABLE IF NOT EXISTS transformdb.dimperson (
  personkey  BIGINT,
  firstname  STRING,
  lastname   STRING
)
CLUSTERED BY (firstname, lastname,personkey) INTO 1 BUCKETS
STORED AS orc
TBLPROPERTIES('transactional' = 'true','orc.compress'='ZLIB','orc.create.index'='true');

CREATE TABLE IF NOT EXISTS transformdb.dimperson001 (
  firstname  STRING,
  lastname   STRING
)
CLUSTERED BY (firstname, lastname) INTO 1 BUCKETS
STORED AS orc
TBLPROPERTIES('transactional' = 'true','orc.compress'='ZLIB','orc.create.index'='true');

INSERT INTO TABLE transformdb.dimperson001
SELECT DISTINCT
  firstname,
  lastname
FROM
  processdb.personhub;

CREATE TABLE IF NOT EXISTS transformdb.dimperson002 (
  personkey  BIGINT,
  firstname  STRING,
  lastname   STRING
)
CLUSTERED BY (firstname, lastname,personkey) INTO 1 BUCKETS
STORED AS orc
TBLPROPERTIES('transactional' = 'true','orc.compress'='ZLIB','orc.create.index'='true');

INSERT INTO TABLE transformdb.dimperson002
SELECT
  ROW_NUMBER() OVER (ORDER BY firstname, lastname),
  firstname,
  lastname
FROM
  transformdb.dimperson001;

INSERT INTO TABLE transformdb.dimperson
SELECT
  personkey,
  firstname,
  lastname
```

```
FROM
  transformdb.dimperson002
ORDER BY firstname, lastname, personkey;

INSERT INTO TABLE transformdb.dimperson
VALUES
(999997,'Ruff','Hond'),
(999998,'Robbie','Rot'),
(999999,'Helen','Kat');

DROP TABLE transformdb.dimperson001;
DROP TABLE transformdb.dimperson002;
```

---

■ **Note**　See example script `Transform02.txt` for the Hive code. It creates and populates the dimension `dimaccount`.

---

```
USE transformdb;

CREATE TABLE IF NOT EXISTS transformdb.dimaccount (
  accountkey       BIGINT,
  accountnumber    INT
)
CLUSTERED BY (accountnumber,accountkey) INTO 1 BUCKETS
STORED AS orc
TBLPROPERTIES('transactional' = 'true','orc.compress'='ZLIB','orc.create.index'='true');

CREATE TABLE IF NOT EXISTS transformdb.dimaccount001 (
  accountnumber    INT
)
CLUSTERED BY (accountnumber) INTO 1 BUCKETS
STORED AS orc
TBLPROPERTIES('transactional' = 'true','orc.compress'='ZLIB','orc.create.index'='true');

INSERT INTO TABLE transformdb.dimaccount001
SELECT DISTINCT
  objectid
FROM
  processdb.objecthub
WHERE objecttype = 'intangible'
AND objectname = 'bankaccount';

CREATE TABLE IF NOT EXISTS transformdb.dimaccount002 (
  accountkey       BIGINT,
  accountnumber    INT
)
CLUSTERED BY (accountnumber,accountkey) INTO 1 BUCKETS
STORED AS orc
TBLPROPERTIES('transactional' = 'true','orc.compress'='ZLIB','orc.create.index'='true');

INSERT INTO TABLE transformdb.dimaccount002
```

```
SELECT DISTINCT
  ROW_NUMBER() OVER (ORDER BY accountnumber DESC),
  accountnumber
FROM
  transformdb.dimaccount001;

INSERT INTO TABLE transformdb.dimaccount
SELECT DISTINCT
  accountkey,
  accountnumber
FROM
  transformdb.dimaccount002
ORDER BY accountnumber;

INSERT INTO TABLE transformdb.dimaccount
VALUES
(88888887,208887),
(88888888,208888),
(88888889,208889);

DROP TABLE transformdb.dimaccount001;
DROP TABLE transformdb.dimaccount002;
```

---

■ **Note**    See example script `Transform03.txt` for the Hive code. It creates and populates the fact
fctpersonaccount.

---

```
USE transformdb;

CREATE TABLE IF NOT EXISTS transformdb.fctpersonaccount (
  personaccountkey      BIGINT,
  personkey             BIGINT,
  accountkey            BIGINT,
  balance               DECIMAL(18, 9)
)
CLUSTERED BY (personkey,accountkey) INTO 1 BUCKETS
STORED AS orc
TBLPROPERTIES('transactional' = 'true','orc.compress'='ZLIB','orc.create.index'='true');

CREATE TABLE IF NOT EXISTS transformdb.fctpersonaccount001 (
  personkey             BIGINT,
  accountkey            BIGINT,
  balance               DECIMAL(18, 9)
)
CLUSTERED BY (personkey,accountkey) INTO 1 BUCKETS
STORED AS orc
TBLPROPERTIES('transactional' = 'true','orc.compress'='ZLIB','orc.create.index'='true');

INSERT INTO TABLE transformdb.fctpersonaccount001
VALUES
(999997,88888887,10.60),
```

```
(999997,88888887,400.70),
(999997,88888887,-210.90),
(999998,88888888,1000.00),
(999998,88888888,1990.60),
(999998,88888888,900.70),
(999999,88888889,160.60),
(999999,88888889,180.70),
(999999,88888889,100.60),
(999999,88888889,120.90),
(999999,88888889,180.69),
(999999,88888889,130.30);

CREATE TABLE IF NOT EXISTS transformdb.fctpersonaccount002 (
  personkey      BIGINT,
  accountkey     BIGINT,
  balance        DECIMAL(18, 9)
)
CLUSTERED BY (personkey,accountkey) INTO 1 BUCKETS
STORED AS orc
TBLPROPERTIES('transactional' = 'true','orc.compress'='ZLIB','orc.create.index'='true');

INSERT INTO TABLE transformdb.fctpersonaccount002
SELECT
CAST(personkey AS BIGINT),
CAST(accountkey AS BIGINT),
CAST(SUM(balance) AS DECIMAL(18, 9))
FROM transformdb.fctpersonaccount001
GROUP BY personkey, accountkey;

INSERT INTO TABLE transformdb.fctpersonaccount
SELECT
ROW_NUMBER() OVER (ORDER BY personkey, accountkey),
CAST(personkey AS BIGINT),
CAST(accountkey AS BIGINT),
CAST(balance AS DECIMAL(18, 9))
FROM transformdb.fctpersonaccount002;

DROP TABLE transformdb.fctpersonaccount001;
DROP TABLE transformdb.fctpersonaccount002;
```

---

■ **Note**    See example script `Transform04.txt` for the Hive code. It creates and populates `dimaddress`, `dimdatetime`, and `fctpersonaddressdate`.

---

```
USE transformdb;

DROP TABLE transformdb.dimaddress;

CREATE TABLE IF NOT EXISTS transformdb.dimaddress(
  addresskey     BIGINT,
  postcode       STRING
```

```
)
CLUSTERED BY (addresskey) INTO 1 BUCKETS
STORED AS orc
TBLPROPERTIES('transactional' = 'true','orc.compress'='ZLIB','orc.create.index'='true');

INSERT INTO TABLE transformdb.dimaddress
VALUES
(1,'KA12 8RR'),
(2,'FK8 1EJ'),
(3,'EH1 2NG');

DROP TABLE transformdb.dimdatetime;

CREATE TABLE IF NOT EXISTS transformdb.dimdatetime(
   datetimekey      BIGINT,
   datetimestr      STRING
)
CLUSTERED BY (datetimekey) INTO 1 BUCKETS
STORED AS orc
TBLPROPERTIES('transactional' = 'true','orc.compress'='ZLIB','orc.create.index'='true');

INSERT INTO TABLE transformdb.dimdatetime
VALUES
(1,'2015/08/23 16h00'),
(2,'2015/10/03 17h00'),
(3,'2015/11/12 06h00');

CREATE TABLE IF NOT EXISTS transformdb.fctpersonaddressdate(
   personaddressdatekey      BIGINT,
   personkey                 BIGINT,
   addresskey                BIGINT,
   datetimekey               BIGINT
)
CLUSTERED BY (datetimekey) INTO 1 BUCKETS
STORED AS orc
TBLPROPERTIES('transactional' = 'true','orc.compress'='ZLIB','orc.create.index'='true');

INSERT INTO TABLE transformdb.fctpersonaddressdate
VALUES
(1,999997,1,1),
(2,999998,2,2),
(3,999999,3,3);
```

If all the scripts completed, check to see that you have all your dimensions and facts, and then execute:

```
SHOW TABLES;
```

You have just completed the transform layer.

# What Have You Mastered

You have successfully created a data warehouse, which includes:

- Creating dimensions.
- Creating facts
- Creating aggregations.

You are making excellent progress. You have mastered the process of building a data warehouse. The hard work is done.

---

■ **Note** Building the data warehouse from the data sources normally takes 70 to 80% of the programming effort in the project.

---

The next phase is to create data marts from your fully functional data warehouse.

# Organize Database

The organize database holds a series of smaller ROLAP (Relational Online Analytical Processing) models consisting of subdivisions of the dimensional and fact model, as described by the sun models, but filtered to create data marts.

You create a database called organisedb to hold the data mart structures.

```
CREATE DATABASE IF NOT EXISTS organisedb;
```

Remember the command you can use in Hive to create the table from another table as a reference.

This works perfectly for data marts, as they contain the same data structure and only have the filtered data from the original table.

```
CREATE TABLE IF NOT EXISTS organisedb.dimperson LIKE transformdb.dimperson;

CREATE TABLE IF NOT EXISTS organisedb.dimaccount LIKE transformdb.dimaccount;

CREATE TABLE IF NOT EXISTS organisedb.fctpersonaccount LIKE transformdb.fctpersonaccount;

CREATE TABLE IF NOT EXISTS organisedb.dimaddress(
  addresskey     BIGINT,
  postcode       STRING
)
CLUSTERED BY (addresskey) INTO 1 BUCKETS
STORED AS orc
TBLPROPERTIES('transactional' = 'true','orc.compress'='ZLIB','orc.create.index'='true');

CREATE TABLE IF NOT EXISTS organisedb.fctpersonaddressdate(
  personaddressdatekey     BIGINT,
  personkey                BIGINT,
  addresskey               BIGINT,
  datetimekey              BIGINT
)
```

```
CLUSTERED BY (datetimekey) INTO 1 BUCKETS
STORED AS orc
TBLPROPERTIES('transactional' = 'true','orc.compress'='ZLIB','orc.create.index'='true');
```

■ **Note** See example script Organise01.txt for the Hive code. It creates and populates the complete organise database.

```
DROP DATABASE organisedb CASCADE;

CREATE DATABASE IF NOT EXISTS organisedb;

USE organisedb;

CREATE TABLE IF NOT EXISTS organisedb.dimperson (
  personkey  BIGINT,
  firstname  STRING,
  lastname   STRING
)
CLUSTERED BY (firstname, lastname,personkey) INTO 1 BUCKETS
STORED AS orc
TBLPROPERTIES('transactional' = 'true','orc.compress'='ZLIB','orc.create.index'='true');

CREATE TABLE IF NOT EXISTS organisedb.dimperson LIKE transformdb.dimperson;

INSERT INTO TABLE organisedb.dimperson
SELECT
  personkey,
  firstname,
  lastname
FROM
  transformdb.dimperson
ORDER BY firstname, lastname, personkey;

CREATE TABLE IF NOT EXISTS organisedb.dimaccount (
  accountkey     BIGINT,
  accountnumber  INT
)
CLUSTERED BY (accountnumber,accountkey) INTO 1 BUCKETS
STORED AS orc
TBLPROPERTIES('transactional' = 'true','orc.compress'='ZLIB','orc.create.index'='true');

CREATE TABLE IF NOT EXISTS organisedb.dimaccount LIKE transformdb.dimaccount;

INSERT INTO TABLE organisedb.dimaccount
SELECT DISTINCT
  accountkey,
  accountnumber
FROM
  transformdb.dimaccount
```

210

```
ORDER BY accountnumber;

CREATE TABLE IF NOT EXISTS organisedb.fctpersonaccount (
  personaccountkey      BIGINT,
  personkey             BIGINT,
  accountkey            BIGINT,
  balance               DECIMAL(18, 9)
)
CLUSTERED BY (personkey,accountkey) INTO 1 BUCKETS
STORED AS orc
TBLPROPERTIES('transactional' = 'true','orc.compress'='ZLIB','orc.create.index'='true');

CREATE TABLE IF NOT EXISTS organisedb.fctpersonaccount LIKE transformdb.fctpersonaccount;
```

Now we create the data marts. We want to select only the record for a specific account holder. Here is the Hive code:

```
INSERT INTO TABLE organisedb.fctpersonaccount
SELECT DISTINCT
  personaccountkey,
  personkey,
  accountkey,
  balance
FROM
  transformdb.fctpersonaccount
WHERE
  personaccountkey = 1
ORDER BY personaccountkey,personkey,accountkey;
```

---

■ **Note**    The where statement enforces the subset of the data warehouse into a data mart.

---

If you execute the following Hive code:

```
SELECT * FROM organisedb.fctpersonaccount;
```

You should only return one record.

You have just mastered the process of organizing data marts.

Let's create one more data mart for addresses. This time we want to slice by columns into a new data mart.

```
CREATE TABLE IF NOT EXISTS organisedb.dimaddress(
  addresskey      BIGINT,
  postcode        STRING
)
CLUSTERED BY (addresskey) INTO 1 BUCKETS
STORED AS orc
TBLPROPERTIES('transactional' = 'true','orc.compress'='ZLIB','orc.create.index'='true');

INSERT INTO TABLE organisedb.dimaddress
SELECT DISTINCT
  addresskey,
```

```
  postcode
FROM
  transformdb.dimaddress
ORDER BY addresskey;
```

Execute the following Hive code:

```
SELECT * FROM organisedb.dimaddress;
```

You have just successfully created a data mart by subselecting specific columns that are important to this data mart.

So lets try an amalgamation of the two requirements.

```
CREATE TABLE IF NOT EXISTS organisedb.fctpersonaddressdate(
  personaddressdatekey       BIGINT,
  personkey                  BIGINT,
  addresskey                 BIGINT,
  datetimekey                BIGINT
)
CLUSTERED BY (datetimekey) INTO 1 BUCKETS
STORED AS orc
TBLPROPERTIES('transactional' = 'true','orc.compress'='ZLIB','orc.create.index'='true');
INSERT INTO TABLE organisedb.fctpersonaddressdate
SELECT
  personaddressdatekey,
  personkey,
  addresskey,
  datetimekey
FROM
  transformdb.fctpersonaddressdate
WHERE personaddressdatekey = 1
ORDER BY
  personaddressdatekey,
  personkey,
  addresskey,
  datetimekey;
```

If the script completed, check that you have all your dimensions and facts for your data marts and execute:

```
SHOW TABLES;
```

Congratulations! You have successfully created a data mart ready to be interrogated for reporting.

## Tips

Subdivide any data warehouse you are moving to a branch server with the data for that branch only. This saves you on network transport and also enhances the speed of the queries in the branch.

Do not do the data mart splitting on the branch servers. Instead, use the more powerful central server and then only transfer the end result of the organise layer with the report layer to the branch. If you can create a separate branch server for the central site, that enables you to process new data without impacting the central branch.

# Report Database

The report database groups the business sun model's results. Create a series of queries to the database to ensure you report consistently across the business. Also create data sets for business entities like the morning report that should stay fixed for the day. There are normally various reports that are created on different intervals, such as hourly, daily, weekly, monthly, quarterly, and yearly.

---

■ **Tip** If you need to create international reports, i.e., produce daily reports at 8h00 local time, use a fixed time for the central processing and add the `timezone` shift in the `organise` layer for the specific branch. That way, your report layer is always set to local time.

---

Let's start:

---

■ **Note** See example script `Report01.txt` for the Hive code. It creates and populates the `report` database.

---

```
DROP DATABASE reportdb CASCADE;

CREATE DATABASE IF NOT EXISTS reportdb;
USE reportdb;

CREATE TABLE IF NOT EXISTS reportdb.report001(
   firstname       STRING,
   lastname        STRING,
   accountnumber   INT,
   balance         DECIMAL(18, 9)
)
CLUSTERED BY (firstname, lastname) INTO 1 BUCKETS
STORED AS orc
TBLPROPERTIES('transactional' = 'true','orc.compress'='ZLIB','orc.create.index'='true');

INSERT INTO TABLE reportdb.report001
SELECT
   dimperson.firstname, dimperson.lastname,
   dimaccount.accountnumber, fctpersonaccount.balance
FROM
   organisedb.fctpersonaccount
JOIN
   organisedb.dimperson
ON
   fctpersonaccount.personkey = dimperson.personkey
JOIN
   organisedb.dimaccount
ON
   fctpersonaccount.accountkey = dimaccount.accountkey;

CREATE TABLE IF NOT EXISTS reportdb.report002(
   accountnumber   INT,
```

```
  last_balance     DECIMAL(18, 9)
)
CLUSTERED BY (firstname, lastname) INTO 1 BUCKETS
STORED AS orc
TBLPROPERTIES('transactional' = 'true','orc.compress'='ZLIB','orc.create.index'='true');

INSERT INTO TABLE reportdb.report002
SELECT
  dimaccount.accountnumber, sum(fctpersonaccount.balance) as last_balance
FROM
  organisedb.fctpersonaccount
JOIN
  organisedb.dimaccount
ON
  fctpersonaccount.accountkey = dimaccount.accountkey;
```

Congratulations! You have completed the Hive data warehouse.

## Example Reports

The data result for Report001 can be reported via a visualization design to convert the data into a business story.

Report all account balances bigger than $998.00.

```
SELECT * FROM reportdb.report001 WHERE balance > 998;
```

This returns 10 results from reportdb.report001.

| Firstname | Lastname | Accountno | Balance |
|-----------|----------|-----------|---------|
| ELISEO | BOULWARE | 68105 | ($1,000.00) |
| SHONNA | HIGBY | 18004 | ($1,000.00) |
| LOUISE | MERINO | 59136 | ($1,000.00) |
| KERSTIN | SAUCEDA | 82385 | ($999.00) |
| NANA | BEHLING | 30073 | ($999.00) |
| SHARDA | DIALS | 18946 | ($1,000.00) |
| VALARIE | BLANKENSHIP | 58597 | ($1,000.00) |
| JAZMINE | HUNSAKER | 69942 | ($999.00) |
| KENNETH | KURTZ | 30669 | ($999.00) |
| DELL | HAWKS | 48440 | ($999.00) |

The data can be formatted using various graphical packages. For example, you could format it as a pie graph (see Figure 8-21).

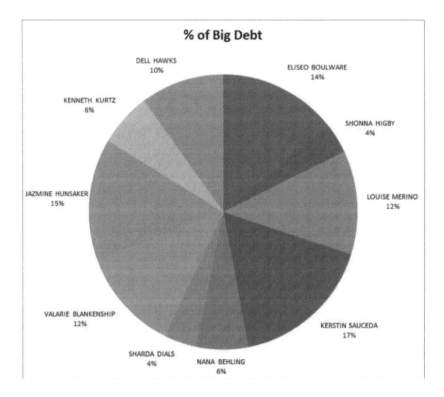

***Figure 8-21.*** *Pie graph*

Or as a bar graph (see Figure 8-22).

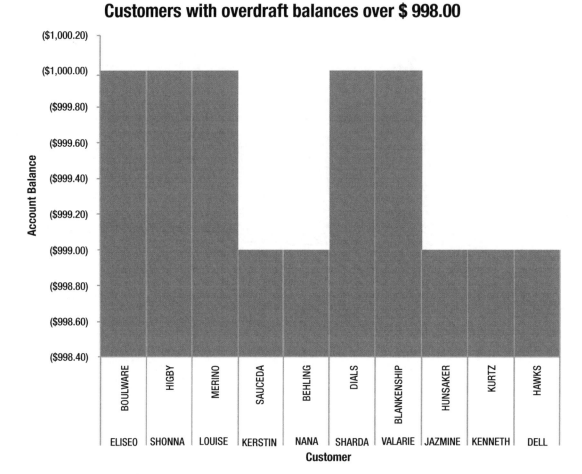

**Figure 8-22.** *Bar graph*

# Advanced Analytics

There are several advanced analytics programs around that enhance the Hive ecosystem. This section covers the integration with R as it supplies an easy open source Hive access route in the analytics environment.

Notable packages are:

- Package hive—R integration of the core of Hadoop and Hive is possible using the correct package (`https://cran.r-project.org/web/packages/hive/hive.pdf`).

- Package NexR RHive 2.0—RHive is an R extension facilitating distributed computing via a Hive query. RHive allows easy usage of HQL (Hive SQL) in R, and allows easy usage of R objects and R functions in Hive (`https://github.com/nexr/RHive`). The user guide is available at `https://github.com/nexr/RHive/wiki/User-Guide`.

# What's Next?

There are many more tools available for Hive, so we suggest you pick your favorite virtualization tool and you will find a Hive connector for that platform. You have completed Chapter 8 and should now:

- Understand the basic data warehouse components:

    - Dimensions with types.

    - Facts and measures—calculated and factless.

- Know how to create sun models for business requirements.

- Convert sun models into star schemas.

- Convert star schemas into Hive code using the Retrieve-Assess-Process-Transform-Organize-Report design principle.

- Understand the construction of the following analytic data structures in Hive.

    - Retrieve—Data imports from external sources.

    - Assess—Enhance data quality.

    - Process—Create a data vault.

    - Transform—Create data warehouse.

    - Organize—Create data marts.

    - Report—Create reports.

Now that you can build a data warehouse and the analytics, proceed to Chapter 9 to master the skills required to secure your data in Hive.

# CHAPTER 9

▓ ▓ ▓

# Performance Tuning: Hive

One of the biggest challenges Hive users face is the slow response time experienced by end users who are running ad hoc queries. When compared to the performance achieved by traditional relation database queries, Hive's response times are often unacceptably slow and often leave you wondering how you can achieve the type of performance your end users are accustomed to.

This chapter presents a systematic approach to diagnosing and improving the performance of Hive queries, which can be easily applied to the majority of your existing Hive tables. Each technique is applied in a cumulative manner thereby compounding the effect. Throughout the process, we will reduce the execution time of a single Hive query from 475 seconds to just under 49 seconds.

## Hive Performance Checklist

In the first part of this chapter, we examine the effect of various optimization techniques against the same query, to better illustrate the impact of each. The cluster used for this testing consists of a single master node with eight cores and 32 GB of RAM, and six worker nodes each with four cores and 32 GB of RAM with Hive version 1.2.1.2.3 installed. The baseline query shown here finds the top five airports that have had the most flights delayed by more than 15 minutes, where the wind speed at the origin airport was above 1 meter/second.

```
SELECT  origin, COUNT(*) as cnt
  FROM flights f JOIN airports a ON (f.origin = a.code)
                JOIN weather w ON (a.station = w.station AND w.year = f.
            year AND w.month = f.month and w.day=f.day)
  WHERE f.depdelay>15 and w.metric = 'AWND' and w.value>10
  GROUP by origin SORT BY cnt DESC LIMIT 5;
```

The data used for the query comes from the three following publicly available data sources that you can download yourself and use to follow along throughout this chapter.

The "flight" data comes from `http://stat-computing.org/dataexpo/2009/the-data.html` and contains flight delay data from 1987-2008 for all U.S. airports. It consists of a total of 123,534,969 rows, each with 29 columns.

The "airport" data contains basic information about all the airports in the United States and can be used to connect the airport code to the weather data. It consists of 3404 rows, each with six data columns, and can be downloaded from `http://stat-computing.org/dataexpo/2009/airports.csv`.

The "weather" data comes from the NOAA web site of historical data, which can be downloaded from `ftp://ftp.ncdc.noaa.gov/pub/data/ghcn/daily/by_year/$year.csv.gz` on a year-by-year basis. For this exercise, we downloaded all of the data for the years 1987 thru 2008 inclusive, which resulted in a total of 636,511,075 rows with 11 data columns each.

© Scott Shaw, Andreas François Vermeulen, Ankur Gupta, David Kjerrumgaard 2016
S. Shaw et al., *Practical Hive*, DOI 10.1007/978-1-4842-0271-5_9

# Execution Engines

Hive currently supports three execution engines, each with its own relative strengths and weaknesses. It is worth noting that while there is a default execution engine for Hive, which is controlled by the hive.execution.engine property in the hive-site.xml file, it is also possible to override this setting on a per-query basis by changing the value of the property at runtime. We will compare the performance of the MapReduce and Tez execution engines next by running the same query using both engines and measuring the performance of each.

## MapReduce

The MapReduce execution engine runs the Hive query as a traditional MapReduce job. It is the original execution engine, and it is also the safest fallback option if your query fails to execute with one of the other execution engines. You can select this execution engine by setting the value of the hive.execution.engine property to mr, i.e., hive.execution.engine=mr. For purposes of this exercise, we will execute the query using the MapReduce execution engine and use this performance as a baseline for our performance improvements. The output from this query shown shows that the query took 475.732 seconds to execute and wrote over 711 MB of intermediate data to the disk in process.

```
MapReduce Jobs Launched:
Stage-Stage-11: Map: 6    Cumulative CPU: 233.33 sec    HDFS Read: 164317688 HDFS Write:
711087924 SUCCESS
Stage-Stage-2: Map: 13   Reduce: 50    Cumulative CPU: 1438.11 sec    HDFS Read: 3278981109
HDFS Write: 268969 SUCCESS
Stage-Stage-3: Map: 4   Reduce: 1    Cumulative CPU: 15.57 sec    HDFS Read: 292269 HDFS Write:
5887 SUCCESS
Stage-Stage-4: Map: 1   Reduce: 1    Cumulative CPU: 3.89 sec    HDFS Read: 10052 HDFS Write:
221 SUCCESS
Stage-Stage-5: Map: 1   Reduce: 1    Cumulative CPU: 4.05 sec    HDFS Read: 4787 HDFS Write: 57
SUCCESS
Total MapReduce CPU Time Spent: 28 minutes 14 seconds 950 msec
OK
ORD      1297377
ATL      1112511
DFW      933903
LAX      626875
PHX      584062
Time taken: 475.732 seconds, Fetched: 5 row(s)
```

## Tez

Apache Tez provides more efficient processing than the MapReduce execution engine, by reducing operations and limiting the amount of intermediate data that is written to disk, as depicted in Figure 9-1. As you can see, the traditional MapReduce execution engine has several steps in which the intermediate data from the reducers are written back to HDFS, which incurs the performance penalty for disk I/O. Contrast this with the data flow of the Tez execution engine shown on the right side, where the reducer's intermediate data is passed directly to the next reducer in the execution plan and bypasses the expense of writing the data to disk.

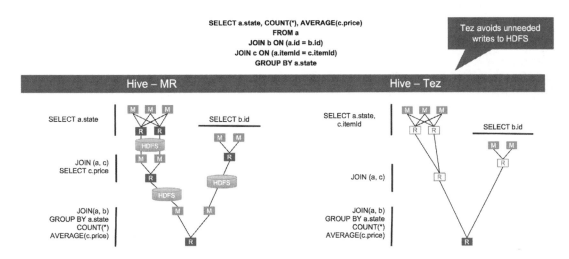

**Figure 9-1.** *Execution engine comparison*

Let's measure the performance of this execution engine by setting the value of the hive.execution.
engine property to tez, i.e., hive.execution.engine=tez, and changing the following two properties
mentioned in Table 9-1—hive.prewarm.enabled =true and hive.prewarm.numcontainers=10. Then we
rerun the query.

```
set hive.execution.engine=tez;
set hive.prewarm.enabled=true;
set hive.prewarm.numcontainers=10;
Total jobs = 1
Launching Job 1 out of 1

Status: Running (Executing on YARN cluster with App id application_1457719973622_0118)

-------------------------------------------------------------------------------
VERTICES      STATUS  TOTAL  COMPLETED  RUNNING  PENDING  FAILED  KILLED
-------------------------------------------------------------------------------
Map 1 ..... SUCCEEDED    22         22        0        0       0       0
Map 5 ..... SUCCEEDED     1          1        0        0       0       0
Map 6 ..... SUCCEEDED    29         29        0        0       0       0
Reducer 2 ..SUCCEEDED    28         28        0        0       0       0
Reducer 3 ..SUCCEEDED    14         14        0        0       0       0
Reducer 4 ..SUCCEEDED     1          1        0        0       0       0
-------------------------------------------------------------------------------
VERTICES: 06/06  [==========================>>] 100%  ELAPSED TIME: 141.23 s
-------------------------------------------------------------------------------
```

```
OK
ORD     1297377
ATL     1112511
DFW     933903
LAX     626875
PHX     584062
Time taken: 166.448 seconds, Fetched: 5 row(s)
```

As you can see, just changing the execution engine resulted in a decrease of the execution time by 309 seconds, or almost 65%. In order to maximize the benefits of the Tez execution engine, you will also want to adjust the configuration settings listed in Table 9-1.

***Table 9-1.*** *Tez-Related Configuration Settings*

| Property | Value | Purpose |
|---|---|---|
| Heap size for HiveServer | 16 GB | Increase the memory from the default of 1 GB. |
| hive.prewarm.enabled | True | Tells Hive to create Tez containers. |
| hive.prewarm.numcontainers | Varies | Tune the number of containers to be held exclusively for Tez. |
| TEZ_CONTAINER_MAX_JAVA_HEAP_ FRACTION | 0.8 | Tez container size is a multiple of YARN container size. |
| hive.auto.convert.join. nonconditionaltask.size | Varies | Tune the map join size. |

# Storage Formats

There are some file formats that are optimized for Hive use, including Parquet and ORC files. Both of these formats are designed to reduce the amount of data read from disk during a query and thereby improve the overall performance of the query.

## The Optimized Row Columnar (ORC) Format

The ORC format is a column-based storage format, meaning that rather than storing all of the data for an individual row of data consecutively on disk, the data for each column of storage contiguously instead. As you can see in Figure 9-2, this allows you to avoid unnecessary disk access for queries that do not contain certain columns, by "skipping over" large sections of data not needed in the results.

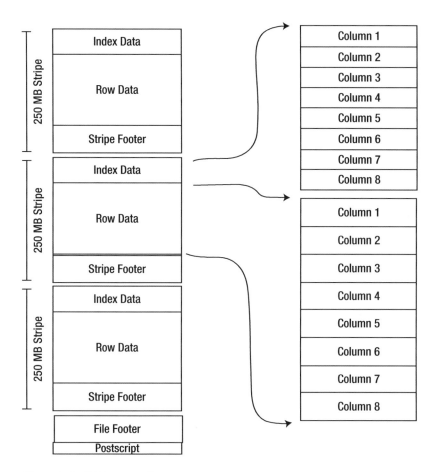

**Figure 9-2.** *ORC storage format*

The ORC format is a splitable file format, meaning that an individual file can be split into multiple block-sized pieces that can be processed in parallel. Each individual block of data is further broken down into 256 MB "stripes" of data that are used to store column data together. Any query that doesn't require that particular column value can "skip" that stride entirely. The ORC format also retains a built-in index, min/max, and other metadata about the contents of each strip in a separate "index data" section of the strip, which allows for fast filtering of stripes based on the query filter parameters.

In order to measure the performance impact of ORC, we must first create two copies of the original tables that will be stored in the ORC format. The quickest way to accomplish this is to run the following CREATE TABLE AS SELECT (CTAS) statements. Then we will modify the query to use the newly created tables and execute.

```
CREATE TABLE flights_orc STORED AS ORC tblproperties("orc.compress"="SNAPPY")
    AS SELECT * FROM flights;
CREATE TABLE weather_orc STORED AS ORC tblproperties("orc.compress"="SNAPPY")
    AS SELECT * FROM weather;
```

```
SELECT  origin, COUNT(*) as cnt
FROM flights_orc f JOIN airports a ON (f.origin = a.code) JOIN weather_orc w ON (a.station =
w.station AND w.year = f.year AND w.month = f.month and w.day=f.day)
WHERE f.depdelay>15 and w.metric = 'AWND' and w.value>10
GROUP by origin SORT BY cnt DESC LIMIT 5;


Total jobs = 1
Launching Job 1 out of 1


Status: Running (Executing on YARN cluster with App id application_1457719973622_0119)


--------------------------------------------------------------------------------
VERTICES        STATUS  TOTAL  COMPLETED  RUNNING  PENDING  FAILED  KILLED
--------------------------------------------------------------------------------
Map 1 ....      SUCCEEDED    22       22       0        0        0       0
Map 5 ....      SUCCEEDED     1        1       0        0        0       0
Map 6 ....      SUCCEEDED    29       29       0        0        0       0
Reducer 2 ...SUCCEEDED       28       28       0        0        0       0
Reducer 3 ...SUCCEEDED       14       14       0        0        0       0
Reducer 4 ...SUCCEEDED        1        1       0        0        0       0
--------------------------------------------------------------------------------
VERTICES: 06/06  [=========================>>] 100%  ELAPSED TIME: 61.60 s
--------------------------------------------------------------------------------
OK
ORD     1297377
ATL     1112511
DFW     933903
LAX     626875
PHX     584062
Time taken: 66.664 seconds, Fetched: 5 row(s)
```

As you can see, utilizing the ORC storage format resulted in a decrease of the execution time by 100 seconds, which is a reduction of over 60%. In order to maximize the benefits of the ORC storage format, you may also want to adjust the following configuration settings when you create a table.

| Property | Value | Notes |
|---|---|---|
| orc.compress | SNAPPY | High-level compression (one of NONE, ZLIB, SNAPPY) |
| orc.compress.size | 262,144 | Number of bytes in each compression chunk |
| orc.stripe.size | 64 MB | Number of bytes in each stripe |
| orc.row.index.stride | 10,000 | Number of rows between index entries (must be >= 1000) |
| orc.create.index | True | Whether to create row indexes or not |

## The Parquet Format

The Parquet format is another column-based storage format that also stores all of the data for each column contiguously on disk, and therefore enjoys performance benefits similar to that of ORC. In order to measure the exact performance impact of Parquet, we must first create two copies of the original tables that will be stored in the Parquet format. The quickest way to accomplish this is to run the following CREATE TABLE AS SELECT (CTAS) statements. Then we will modify the query to use the newly created tables and execute.

```
CREATE TABLE flights_parquet STORED AS Parquet AS SELECT * FROM flights;
CREATE TABLE weather_parquet STORED AS Parquet AS SELECT * FROM weather;

SELECT  origin, COUNT(*) as cnt
FROM flights_parquet f JOIN airports a ON (f.origin = a.code) JOIN weather_parquet w ON
(a.station = w.station AND w.year = f.year AND w.month = f.month and w.day=f.day)
WHERE f.depdelay>15 and w.metric = 'AWND' and w.value>10
GROUP by origin SORT BY cnt DESC LIMIT 5;

Launching Job 1 out of 1

Status: Running (Executing on YARN cluster with App id application_1457719973622_0121)
```

```
--------------------------------------------------------------------------------
VERTICES       STATUS   TOTAL  COMPLETED  RUNNING  PENDING  FAILED  KILLED
--------------------------------------------------------------------------------
Map 1 ....... SUCCEEDED     67        67        0        0       0       0
Map 5 ........SUCCEEDED      1         1        0        0       0       0
Map 6 ........SUCCEEDED     60        60        0        0       0       0
Reducer 2 ....SUCCEEDED      1         1        0        0       0       0
Reducer 3 ....SUCCEEDED      1         1        0        0       0       0
Reducer 4 ....SUCCEEDED      1         1        0        0       0       0
--------------------------------------------------------------------------------
VERTICES: 06/06  [==========================>>] 100%  ELAPSED TIME: 112.39 s
--------------------------------------------------------------------------------
OK
ORD    1297377
ATL    1112511
DFW    933903
LAX    626875
PHX    584062
Time taken: 113.938 seconds, Fetched: 5 row(s)
```

The Parquet storage format resulted in a decrease of the execution time by 53 seconds, which is a reduction of almost 32%. While this is an improvement over using just the Tez execution engine, it is still not as good of an improvement as seen with the ORC format.

# Vectorized Query Execution

Hive's default query execution engine processes one row at a time that requires multiple layers of virtual method calls within the nested loop, which is very inefficient from a CPU perspective. Vectorized query execution is a Hive feature that aims to eliminate these inefficiencies by reading the rows in batches of 1024 and applying the operation on the entire collection of records at a time rather than individually. This vector mode of execution has been proven to be an order of magnitude faster for typical query operations such as scans, filters, aggregates, and joins. In order to use vectorized query execution, you must store your data in ORC format.

Let's measure the performance of this execution engine by setting the value of the hive.vectorized. execution.enabled property to true and running the query against the ORC backed tables.

```
set hive.vectorized.execution.enabled = true;

SELECT  origin, COUNT(*) as cnt
FROM flights_orc f JOIN airports a ON (f.origin = a.code) JOIN weather_orc w ON (a.station =
w.station AND w.year = f.year AND w.month = f.month and w.day=f.day)
WHERE f.depdelay>15 and w.metric = 'AWND' and w.value>10
GROUP by origin SORT BY cnt DESC LIMIT 5;

Launching Job 1 out of 1

Status: Running (Executing on YARN cluster with App id application_1457719973622_0122)
```

```
--------------------------------------------------------------------------------
VERTICES        STATUS  TOTAL  COMPLETED  RUNNING  PENDING  FAILED  KILLED
--------------------------------------------------------------------------------
Map 1 .......SUCCEEDED     22        22        0        0       0       0
Map 5 .......SUCCEEDED      1         1        0        0       0       0
Map 6 .......SUCCEEDED     29        29        0        0       0       0
Reducer 2 ...SUCCEEDED     28        28        0        0       0       0
Reducer 3 ...SUCCEEDED     14        14        0        0       0       0
Reducer 4 ...SUCCEEDED      1         1        0        0       0       0
--------------------------------------------------------------------------------
VERTICES: 06/06  [==========================>>] 100%  ELAPSED TIME: 50.60 s
--------------------------------------------------------------------------------
OK
ORD     1297377
ATL     1112511
DFW     933903
LAX     626875
PHX     584062
Time taken: 52.174 seconds, Fetched: 5 row(s)
```

The vectorized query execution resulted in a decrease of the execution time by 12 seconds over just Tez and ORC alone, which is a reduction of over 18%.

# Query Execution Plan

The Hive driver is responsible for translating the SQL statement into an execution plan for the target execution engine by following the sequence depicted in Figure 9-3:

1. The *parser* parses the SQL statement and produces an abstract syntax tree (AST) that represents the logical operations that must be performed in order to generate the correct result set, e.g., SELECTs, JOINs, UNIONs, groupings, projections, and so on.

2. The *planner* takes the AST and retrieves table metadata from the Hive Metastore, including the HDFS file location, storage formats, number of rows, and so on.

3. The *query optimizer* uses the AST and table metadata from the previous steps and produces a physical operation tree known as the execution plan that represents all the physical operations that must be performed to retrieve the data, e.g., a nested loop join, sort-merge join, hash join, index join, and so on.

The execution plan generated by the query optimizer ultimately determines the tasks that will be executed on your Hadoop cluster. Consequently, they have the biggest performance impact in a data analytics system such as Hive, since the difference between generating the right or wrong execution plan could mean seconds, minutes, or even hours of additional execution time.

The CBO helps the Hive driver produce an optimal execution plan by leveraging the table statistics in order to make informed decisions on performance costs of each possible execution plan it generates.

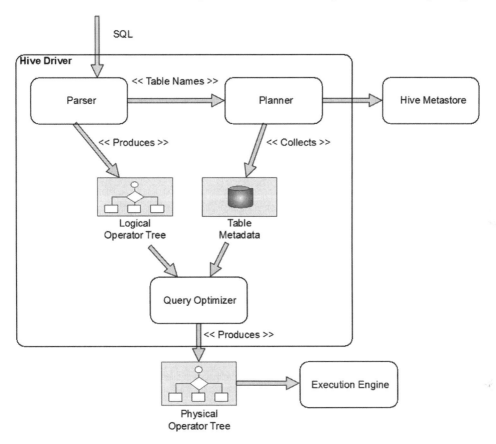

**Figure 9-3.** *Hive driver execution flow*

## Cost-Based Optimization

The cost-based optimization (CBO) engine uses statistics in the Hive Metastore to produce optimal query plans. There are two types of statistics that are used for optimization: table stats, which include the uncompressed size of the table, number of rows, and number of files used to store the data, and column stats, which include NDV (number of distinct values) and min/max/count values.

The CBO does join reordering, improves plans for star and bushy join schemas, and provides opportunistic improvements based on sample queries. The downside of the CBO is the fact that you must gather and maintain accurate statistics about your tables in order for the cost-based optimization engine to be effective. Unfortunately, the collection of table statistics is an expensive operation, but the benefits can be reaped on all subsequent queries involving the table for which statistics were collected. You can automate the global collection of table statistics, by setting the hive.stats.autogather property to true inside hive-site.xml. Since that was not the value of the property when we first created our ORC backed tables, we will need to issue the following command to gather the table statistics for us:

```
ANALYZE TABLE weather_ORC COMPUTE STATISTICS;
Table weather stats: [numFiles=29, numRows=832252480, totalSize=2600971165,
rawDataSize=242185471680]

ANALYZE TABLE weather_ORC COMPUTE STATISTICS FOR COLUMNS;
```

```
--------------------------------------------------------------------------
VERTICES       STATUS  TOTAL  COMPLETED  RUNNING  PENDING  FAILED  KILLED
--------------------------------------------------------------------------
Map 1 .......SUCCEEDED    29         29        0        0       0       0
Reducer 2 ...SUCCEEDED     1          1        0        0       0       0
--------------------------------------------------------------------------
VERTICES: 02/02  [==========================>>] 100%  ELAPSED TIME: 197.79 s
--------------------------------------------------------------------------
OK
Time taken: 216.449 seconds
```

```
ANALYZE TABLE flights_ORC COMPUTE STATISTICS;
Table flights stats: [numFiles=22, numRows=123534969, totalSize=1632812702,
rawDataSize=73119762912]

ANALYZE TABLE flights_ORC COMPUTE STATISTICS FOR COLUMNS;
```

```
--------------------------------------------------------------------------
VERTICES       STATUS  TOTAL  COMPLETED  RUNNING  PENDING  FAILED  KILLED
--------------------------------------------------------------------------
Map 1 .......SUCCEEDED    22         22        0        0       0       0
Reducer 2 ...SUCCEEDED     1          1        0        0       0       0
--------------------------------------------------------------------------
VERTICES: 02/02  [==========================>>] 100%  ELAPSED TIME: 184.85 s
--------------------------------------------------------------------------
OK
Time taken: 186.767 seconds
```

Once the stats have been computed, we can enable the CBO by setting the following properties inside Hive so that every query we run will now use the cost-based optimization engine.

```
SET hive.cbo.enable=true;
SET hive.compute.query.using.stats = true;
SET hive.stats.fetch.column.stats = true;
SET hive.stats.fetch.partition.stats = true;

SELECT  origin, COUNT(*) as cnt
  FROM flights f JOIN airports a ON (f.origin = a.code)
                 JOIN weather w ON (a.station = w.station AND w.year = f.
             year AND w.month = f.month and w.day=f.day)
  WHERE f.depdelay>15 and w.metric = 'AWND' and w.value>10
  GROUP by origin SORT BY cnt DESC LIMIT 5;
```

```
--------------------------------------------------------------------------------
VERTICES        STATUS  TOTAL  COMPLETED  RUNNING  PENDING  FAILED  KILLED
--------------------------------------------------------------------------------
Map 1 .......SUCCEEDED    22        22        0        0       0       0
Map 5 .......SUCCEEDED     1         1        0        0       0       0
Map 6 .......SUCCEEDED    29        29        0        0       0       0
Reducer 2 ...SUCCEEDED    77        77        0        0       0       0
Reducer 3 ...SUCCEEDED    39        39        0        0       0       0
Reducer 4 ...SUCCEEDED     1         1        0        0       0       0
--------------------------------------------------------------------------------
VERTICES: 06/06  [==========================>>] 100%  ELAPSED TIME: 45.98 s
--------------------------------------------------------------------------------
OK
ORD     1297377
ATL     1112511
DFW     933903
LAX     626875
PHX     584062
Time taken: 48.4 seconds, Fetched: 5 row(s)
```

The CBO engine further reduced the execution time by another four seconds, or 7%, and brings the final optimization. While the CBO's impact wasn't significant, there are other queries in which the impact is much more profound, such as when your JOIN statements aren't in the optimal order. In order to view the execution plan produced by the CBO, you can utilize the Hive EXPLAIN command to display the execution plan, which has the following syntax:

```
EXPLAIN [EXTENDED|DEPENDENCY|AUTHORIZATION] query
```

The EXPLAIN output consists of three parts—the Abstract Syntax Tree for the query, the dependencies between the different stages of the plan, and a description of each of the stages. As an example, consider the following EXPLAIN command and the corresponding execution plan:

```
EXPLAIN
SELECT  origin, COUNT(*) as cnt
  FROM flights f JOIN airports a ON (f.origin = a.code)
                 JOIN weather w ON (a.station = w.station AND w.year = f.
             year AND w.month = f.month and w.day=f.day)
  WHERE f.depdelay>15 and w.metric = 'AWND' and w.value>10
  GROUP by origin SORT BY cnt DESC LIMIT 5;

OK
STAGE DEPENDENCIES:
   Stage-1 is a root stage
   Stage-0 depends on stage 1.

STAGE PLANS:
   Stage: Stage-1
     Tez
       Edges:
          Map 1 <- Map 5 (BROADCAST_EDGE), Map 6 (BROADCAST_EDGE)
          Reducer 2 <- Map 1 (SIMPLE_EDGE)
```

```
        Reducer 3 <- Reducer 2 (SIMPLE_EDGE)
        Reducer 4 <- Reducer 3 (SIMPLE_EDGE)
    DagName: ch08_2016042270101_a64ba841-734b6-3517-8f96-ed7bf89e92b4:2
    Verticies:
        Map 1
            Map Operator Tree
                ......
        Map 5
            Map Operator Tree
                ......
        Map 6
            Map Operator Tree
                ......
```

## The Execution Plan

We will examine each of the map operations in great detail momentarily, but we first want to stop and point out what information we can glean from this portion of the EXPLAIN output. First off, we can see that there are exactly two stages in this execution plan—Stage-1 does all of the work to generate the results and Stage-0 returns the results to the end user and depends on Stage-1. Secondly, we can see that the DAG for Stage-1 is as shown in Figure 9-4.

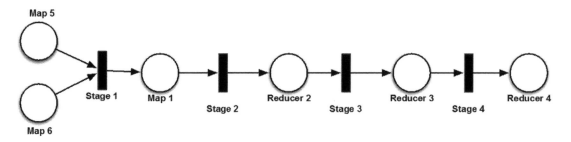

***Figure 9-4.*** *Execution plan DAG*

As we will see next, map phase 5 prepares the weather data for the JOIN operation by applying the filter to the data set to return only those rows that match the criteria in the WHERE clause. Similarly, map phase 6 applies a filter to the airport table before sending it off to map phase 1, which performs the three-way join to connect the weather, the airport, and the flight data. Reducers 3 and 4 perform the COUNT and LIMIT functions before returning the result set to the user. Now let's look at these phases in great detail, starting at the top of the DAG and working our way down the execution chain.

```
Map 5
    Map Operator Tree:
        TableScan
            alias: w
            filterExpr: (((((station is not null and year is not null) and month
            is not null) and day is not null) and (metric='AWND')) and (value >
            10)) (type:boolean)
            Statistics: Num rows: 636511075 Data size: 2592872704 Basic stats:
            COMPLETE Column stats: COMPLETE
```

```
         Filter Operator
         predicate: ((((((station is not null and year is not null) and month
         is not null) and day is not null) and (metric='AWND')) and (value >
         10)) (type:boolean)
         Statistics: Num rows: 1982900 Data size: 394597100 Basic stats:
         COMPLETE Column stats: COMPLETE
         Reduce output Operator
            Key expressions: station (type: string), year (type: int), month(type:
            Int), day(type: int)
            Sort order: ++++
            Map-reduce partition columns: station (type: string), year (type:
            int), month(type: int), day(type: int)
            Statistics: Num rows: 1982900 Data size: 394597100 Basic stats:
            COMPLETE Column stats: COMPLETE
         Execution mode: Vectorized
```

Observations about map phase 5: It is handling the weather table and it is applying a filter based on all four of the tables' partition keys, which helps reduce the number of records processed to just 1,982,900 rows out of the 636,511,075 rows in the table. So instead of having to process 2.6 GB of data, we only have 8 MB to deal with.

```
Map 6
   Map Operator Tree:
      TableScan
        alias: a
        filterExpr: (code is not null and station is not null) (type: boolean)
        Statistics: Num rows: 3404 Data size: 166345 Basic stats: COMPLETE Column
        Stats: NONE
        Filter Operator
           predicate: (code is not null and station is not null)(type: boolean)
           Statistics: Num rows: 851 Data size: 41586 Basic stats: COMPLETE
           Column Stats: NONE
           Reduce Output Operator
              key expressions: code (type: string)
              sort order: +
              Map-reduce partition columns: code (type: string)
              Statistics: Num rows: 851 Data size: 41586 Basic stats: COMPLETE
              Column Stats: NONE
              Value expressions: station (type: string)
```

Observations about map phase 6: It is handling the airport table, which only has 3404 rows to begin with, but this mapper filters it down further to just 851 rows and prepares the data set for the map join, which occurs in Map phase 1.

```
Map 1
   Map Operator Tree:
      TableScan
        alias: f
        filterExpr: ((((origin is not null and year is not null) and month is not
        null) and day is not null) and (depdelay > 15)) (type: boolean)
        Statistics: Num rows: 41178523 Data size: 4238297753 Basic stats:
        COMPLETE Column stats: COMPLETE
```

```
Filter Operator
    predicate: (((origin is not null and year is not null) and month is
    not null) and day is not null) and (depdelay > 15)) (type: boolean)
    Statistics: Num rows: 41178523 Data size: 4238297753 Basic stats:
    COMPLETE Column stats: COMPLETE
    Map Join Operator
        condition map:
            Inner join 0 to 1
        condition expressions:
            0 {year} {month} {day} {origin}
            1 {station}
        keys:
            0 origin (type: string)
            1 code (type: string)
        outputColumnNames: _col0, _col1, _col2, _col16, _col37
        input vertices:
            1 Map 6
        Statistics: Num rows: 4596156 Data size: 4662127629 Basic stats:
        COMPLETE Column stats: NONE
        Map Join Operator
            condition map:
                Inner Join 0 to 1
            Condition expressions:
                0 (_col16}
                1
            keys:
                0 _col37 (type: string), _col0 (type: int), _col1 (type:
                  int), _col2 (type: int)
                1 station (type: string), year (type: int), month (type:
                    int), day (type: int)
            outputColumnNames: _col16
            input vertices:
                1 Map 5
            Statistics: Num rows: 49825772 Data size: 5128340503 Basic
            stats: COMPLETE Column stats: NONE
```

As you can see, the CBO helped generate an optimal execution plan in which the amount of data read from disk and processed was greatly reduced at the earliest possible point in the execution, making the overall job more efficient.

## Performance Checklist Summary

Overall, we were able to reduce the execution time of a single query involving two large tables from 475 seconds to under 49 seconds (almost a 10x improvement) using just a few techniques such as Tez, ORC storage format, vectorized query execution, and the cost-based optimization engine. Best of all, most if not all of these techniques can be applied to the majority of your existing Hive tables with minimal effort.

# Hive Security

Data is one of the most valuable assets of any organization. Loss of information is probably one of the worst nightmares in any organization. Incidents of such nature can cause not only a significant financial loss but can also result in an epoch-making damage to the reputation of the company. Protecting your data asset requires appropriate security solutions in place to avoid breaches. Implementing strong security solutions requires a thorough planning and design stage with a strong need to recognize the risks associated with the platform.

Hadoop is a distributed system for storing and processing large amounts of data in a single shared platform known as a *data lake*. There are many advantages of bringing the data from various systems in a data lake. It allows data scientists to discover various insights by co-relating the data sets that were otherwise stored in various silos. These data sets will still be of interest to various business users, who should only be able to access the data that they are supposed to. In some industries, there are strict rules that drive such distinction of access between various types of users or business units. The organizations that operate in this space often invest significant amount of money to ensure proper controls are in place.

In this chapter we revisit the aspects of data security and discuss the current state of security in Hive. We also visit various types of privileges in Hive, which are maintained using Apache Ranger, the security solution of Apache Hadoop. Finally, we also look at how Apache Ranger maintains an audit record of the data accessed using Hive.

## Data Security Aspects

Before we look into the state of security in Hadoop, lets discuss various aspects of any data security solution, shown in Table 10-1.

*Table 10-1.* *Various Aspects of Security*

| Security Aspect | Feature | Purpose |
| --- | --- | --- |
| Administration | Central management and consistent security | How can I set policy across the entire cluster? |
| Authentication/perimeter security | Authenticate users and systems | Who am I/prove it? |
| Authorization | Provision access to data | What can I do? |
| Audit | Maintain a record of data access | What did I do? |
| Data protection | Protect data at rest and in motion | How can I encrypt data at rest and over the wire? |

© Scott Shaw, Andreas François Vermeulen, Ankur Gupta, David Kjerrumgaard 2016
S. Shaw et al., *Practical Hive*, DOI 10.1007/978-1-4842-0271-5_10

# Authentication

Authentication is a process of verifying someone's identity, i.e., ensuring that someone is who they claim to be. This is done by comparing the credentials provided by the individual or a software service to what is stored on the file or on an authentication server. If the credentials match, the user or the machine is granted access. This is the first step in granting users access to any system. Various authentication systems use different methods to perform authentication. An enterprise system requires the authentication mechanisms to be rigid enough to ensure that the credentials aren't easy to guess or compromised by someone eavesdropping on the network.

# Authorization

Authorization is a way of controlling the resources that can be accessed by a verified user in a system. In a multi-tenanted system, this is perhaps the most critical element of security. Without appropriate authorization system in place, there is no way to control who can access what. Every time an authenticated user makes a request to access a resource, the authorization system uses the access control rules to determine whether that user should be granted access to that resource. These access control rules are created by security administrators.

# Administration

Administration is the process of managing the users of a system. As the number of users grows in any system, this becomes a complex challenge. You can create the most sophisticated security policies in a system, but if they are not applied to the users correctly, the system won't be truly secure. Hence, it is the job of user administrators to ensure that the right policies are not only defined but also correctly applied to various types of users. Most companies often perform a regular check to ensure that the security policies are applied as they should be and there is no deviation from what they should be. That includes, for example, ensuring that users in a particular group do not have access to a particular part of the system that they don't need.

# Auditing

Auditing is a process for maintaining a trail of of every access that was granted or denied to users. The audit trail provides a view of the day-to-day health of a system's security architecture. Looking at the audit trail, administrators can determine who accessed what and if there are any users who tried to access something that they shouldn't have. Maintaining an audit trail is often a legal requirement in many industries and enterprises are required to show these to third-party companies, which perform regular audits of their entire security infrastructure.

# Data Protection

In today's world, data is one of the most critical assets of any enterprise. Different security standards like PCI DSS require this data to be protected. This protection is required both for the data at rest and while it is accessed by someone. There are multiple security protocols that ensure that online data (while it is accessed) is secure. These protocols are widely used by various systems on the Internet. However, the systems also need to ensure that the data stored on the disk is also protected. Even if someone steals the physical disk from a data center, the information stored on the storage media should be protected in a way that it cannot be interpreted.

# Hadoop Security

Since its advent, Hadoop has come a long way both in terms of its functionality and the way you secure data in it. It started as a project to store and index the web in a distributed platform, whereby achieving the performance and other features were much more important than ensuring appropriate security measures were in place. The earlier version of Hadoop relied on querying the OS level parameters to crosscheck the username. These parameters could be very easily set to any values allowing impersonation. However, as Hadoop became popular, more and more companies started to use it to store and process huge data sets in large clusters.

The concept of YARN resulted in the transformation of Hadoop silos into an enterprise data lake, which could be used by various business units to run batch, interactive, and real-time workloads. Lack of security was a huge barrier in the adoption of Hadoop and the community recognized this. The distributed nature of Hadoop makes it difficult to implement security in a cluster. A typical Hadoop cluster consists of many nodes and the interaction between a client process and the actual job process happens at various levels. Many times, the user who submits the job is different from the user who actually executes the code at the processing time. The addition of various processing engines in the Hadoop ecosystem made security in Hadoop even more difficult but more important. Multiple processing engines are executed in a distributed manner and require the authorization checks to be executed at multiple layers. This is now handled by integrating the Hadoop infrastructure with Apache Ranger.

The purpose of this chapter is not to go into too much detail of the history of security options in Hadoop, but to discuss the current state of security. The Apache open source community has put in an enormous amount of effort to integrate the Hadoop stack with standard security solutions like Active Directory, LDAP, and Kerberos for authentication. Authorization of users to various data sets for different processing engines is now done using Apache Ranger. Apache Ranger also provides auditing capabilities in Hadoop. Further, the data stored in Hadoop can be protected using HDFS Transparent Data Encryption and encrypted over the Internet using security protocols like SSL/TCL. We look into more details about these solutions later in this chapter.

# Hive Security

Hive started as a project to write processing jobs using SQL in Hadoop without having to write complex Java for MapReduce. At the time when Hive was written, Hadoop was not integrated with existing enterprise security solutions. Since then a lot has changed, especially in terms of how the authorization access is controlled in Hive. Let's take a look at the various authorization modes in Hive.

## Default Authorization Mode

This is also known as the legacy authorization mode. This was the only authorization model available until Hive version 0.10.0. There were many security vulnerabilities in this mode and hence it was not very well suited to provide a secure environment. In terms of its working it was quite similar to a relational database. Just like a relational database, there was a concept of users/groups and roles. The privileges could be added to a group and the group could then be assigned to a single user(s) or groups. By default, when a user created a table under this mode, no privileges were granted to the person who created the table.

This authorization mode was enabled by modifying the value of `hive.security.authorization.enabled` to `true` in the `hive-site.xml` file, as shown.

```
<property>
    <name>hive.security.authenticator.manager</name>
    <value>org.apache.hadoop.hive.ql.security.ProxyUserAuthenticator</value>
</property>

<property>
    <name>hive.security.authorization.enabled</name>
    <value>true</value>
</property>

<property>
    <name>hive.security.authorization.manager</name>
    <value>org.apache.hadoop.hive.ql.security.authorization.plugin.sqlstd.
    SQLStdConfOnlyAuthorizerFactory</value>
</property>
```

This mode was quite similar to RDBMS style authorization. The access was managed at various levels like users, groups, and roles. This authorization mode also had some properties to control the default privileges that the users, groups, and roles would get when a new table was created.

## Storage-Based Authorization Mode

The storage-based authorization mode was added in later versions of Hive. It relies on permission model of HDFS, the filesystem of Hadoop. In this type of security model, the HDFS permissions act as a single source of truth and Hive simply relies on this single source of truth to determine whether or not the access should be granted to a user request. When a user tries to access a table, Hive checks the permissions of the underlying directories on the filesystem to control the security to the Hive objects.

The storage-based authorization mode can be enabled by setting the following properties in hive-site.html.

```
<property>
    <name>hive.security.metastore.authenticator.manager</name>
    <value>org.apache.hadoop.hive.ql.security.HadoopDefaultMetastoreAuthenticator</value>
</property>

<property>
    <name>hive.security.metastore.authorization.auth.reads</name>
    <value>true</value>
</property>

<property>
    <name>hive.security.metastore.authorization.manager</name>
    <value>org.apache.hadoop.hive.ql.security.authorization.
    StorageBasedAuthorizationProvider</value>
</property>

<property>
    <name>hive.server2.allow.user.substitution</name>
    <value>true</value>
</property>
```

Since Hive CLI is deprecated, most of the users will be connecting to HiveServer2 either by using Beeline or another tool using JDBC/ODBC. It is important to also set another parameter called `hive.server2.enable.doAs` to `true` for this authorization mode to work.

```
<property>
    <name>hive.server2.authentication.spnego.keytab</name>
    <value>HTTP/_HOST@EXAMPLE.COM</value>
</property>

<property>
    <name>hive.server2.authentication.spnego.principal</name>
    <value>/etc/security/keytabs/spnego.service.keytab</value>
</property>

<property>
    <name>hive.server2.enable.doAs</name>
    <value>true</value>
</property>

<property>
    <name>hive.server2.logging.operation.enabled</name>
    <value>true</value>
</property>
```

This parameter determines the end user with which HiveServer2 executes the queries. When it's set to `true`, HiveServer2 executes the queries as the user who was authenticated; otherwise, it uses the user ID with which HiveServer2 process was started, which in most cases is the Hive.

This authorization mode is suitable if the users also require direct access to the data files on HDFS for running other jobs using one of the other processing paradigms like Pig, Spark, or even MapReduce.

---

HDFS ACLs provide a lot of flexibility to manage file-level access. If the users only require access using SQL, then use the SQL standards-based authorization mode.

---

## SQL Standards-Based Authorization Mode

This authorization mode provides a way to control access to a much finer level than storage-based authorization. If the users of Hive are connecting to HiveServer2 and only require access to the data using SQL, this is the recommended authorization mode. In this mode, you can control access at column, view level, as the HiveServer2 API understands the concept of rows and columns. This also provides a mechanism that can be integrated with Apache Ranger to define policies for managing access. We discuss the Hive plug-in for Ranger later in this chapter.

In order to enable this security mode, you need to set the following parameters in `hive-site.xml`.

```
<property>
    <name>hive.security.authorization.manager</name>
    <value>org.apache.hadoop.hive.ql.security.authorization.plugin.sqlstd.
    SQLStdConfOnlyAuthorizerFactory</value>
</property>
```

```
<property>
    <name>hive.server2.doAs</name>
    <value>false</value>
 </property>

  <property>
      <name>hive.security.metastore.authenticator.manager</name>
      <value>org.apache.hadoop.hive.ql.security.HadoopDefaultMetastoreAuthenticator</value>
      </property>

      <property>
        <name>hive.security.metastore.authorization.auth.reads</name>
        <value>true</value>
      </property>
```

The general best practice is to allow users access only through HiveServer2 and to restrict the user code and non-SQL commands that can be run. When a user submits a request, the privileges are checked but the actual query is executed as the Hive server user. You should also lock down the access to the actual data at the HDFS level, by giving the permission only to the Hive server user. If there are any additional users who don't require access through SQL but only need access to these files at the HDFS level, you can create ACLs for them.

## Managing Access through SQL

Just like with standard SQL, you can manage access control in Hive using privileges, users, roles, and objects. Privileges are granted to users and roles. Users belong to one or more roles and they can enable a role. Some of the privileges that can be granted in Hive are ALTER, DROP, INDEX, LOCK, SELECT, INSERT, UPDATE, DELETE, and CREATE, ALL. If you are familiar with standard SQL, you will find that the commands to manage privileges in Hive are quite similar. We now look at some examples for creating and managing privileges for various objects in Hive.

Let's first create a database.

```
CREATE DATABASE TEST;
```

We will now create a new table in the TEST database.

```
USE TEST;
CREATE TABLE TESTING (A INT, B STRING);
```

Grant SELECT privilege on the TESTING table to user JOHN:

```
GRANT SELECT on TABLE TESTING TO USER JOHN;
```

Verify the GRANTS on TABLE TESTING:

```
SHOW GRANT ON TABLE TESTING;
```

Verify all grants for user JOHN:

```
SHOW GRANT USER JOHN ON ALL;
```

You can enable a ROLE for a user using the SET ROLE command.

```
SET ROLE BI_ROLE;
```

The grant can also be provided on column level for a table.

```
GRANT SELECT ON TABLE TESTING COLUMN A TO USER SCOTT;
```

You can enable partition-level privileges for a table and then control the privileges for those partitions.

```
CREATE TABLE TESTING (A INT, B STRING)  PARTITIONED BY (state string);
ALTER TABLE TESTING  SET TBLPROPERTIES ('PARTITION_LEVEL_PRIVILEGE'='TRUE');
GRANT SELECT ON TABLE TESTING PARTITION (state='NY') to USER SCOTT;
```

Loading data into a table requires the UPDATE privilege.

```
GRANT UPDATE on TABLE TESTING TO USER JOHN;
LOAD DATA INPATH '/tmp/hive/testing.csv' into TABLE TESTING;
```

Just like standard SQL, you have the option to grant the privileges with GRANT OPTION and ADMIN OPTION for roles. This allows the recipient of the privilege to grant them to another user.

# Hive Authorization Using Apache Ranger

Apache Ranger is a framework for enabling, monitoring, and managing the comprehensive data security across the Hadoop platform. Ranger simply helps a Hadoop admin with various security management tasks. It provides a mechanism to manage the security from a single pane for various components. With Ranger, you can control fine-grained access to various components of the Hadoop ecosystem. As shown in Figure 10-1, it has a set of built-in plug-ins that integrate with various processing engines, including Hive. When a user runs a query using a client that connects to HiveServer2, the Hive plug-in for Ranger, which is integrated with HiveServer2, evaluates the privileges from the pool of its access policies defined using the Ranger control panel.

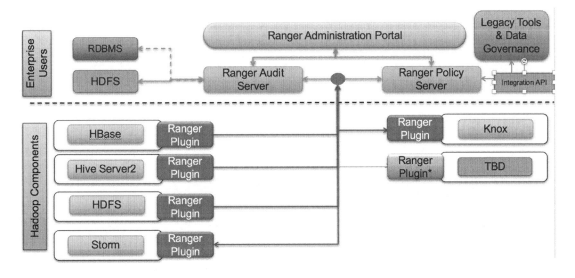

***Figure 10-1.*** *Ranger architecture*

As you can see from the architecture in Figure 10-1, Ranger has an Administration Portal you can use to define various policies for different components. It also has a built-in Policy Server, where all defined policies are maintained. Ranger stores these in a policy database, which is currently deployed in a RDBMS. Ranger also has a built-in Audit Server, which we discuss later in this chapter.

---

■ **Note** The further sections of this chapter assume that you have already installed Ranger in your demo environment. The installation and integration of Ranger with Active Directory/LDAP is beyond the scope of this book. It is documented on Apache Ranger web site and can be accessed through this link: `https://cwiki. apache.org/confluence/display/RANGER/Apache+Ranger+0.5.0+Installation`.

The focus of this chapter is to define the Hive access policies in Ranger and then verify that they are enforced by checking the audit records.

---

## Accessing the Ranger UI

You can access the Ranger UI using the following URL:

`http://rangerserver:6080`

When you log in to the Ranger UI, the home page lists the various menus and types of policies that can be created using Ranger (as shown in Figure 10-2).

*Figure 10-2.* *Ranger user interface*

## Creating Ranger Policies

Use these steps to create a new policy in the Ranger UI:

1. Click on the policy group name under Hive. As shown in Figure 10-3, this should bring up the page with a list of existing Hive policies.

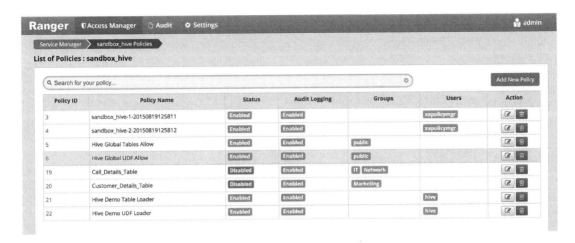

*Figure 10-3. List of policies in Ranger*

2. Now click on Add New Policy to open a new page, similar to the one in Figure 10-4.

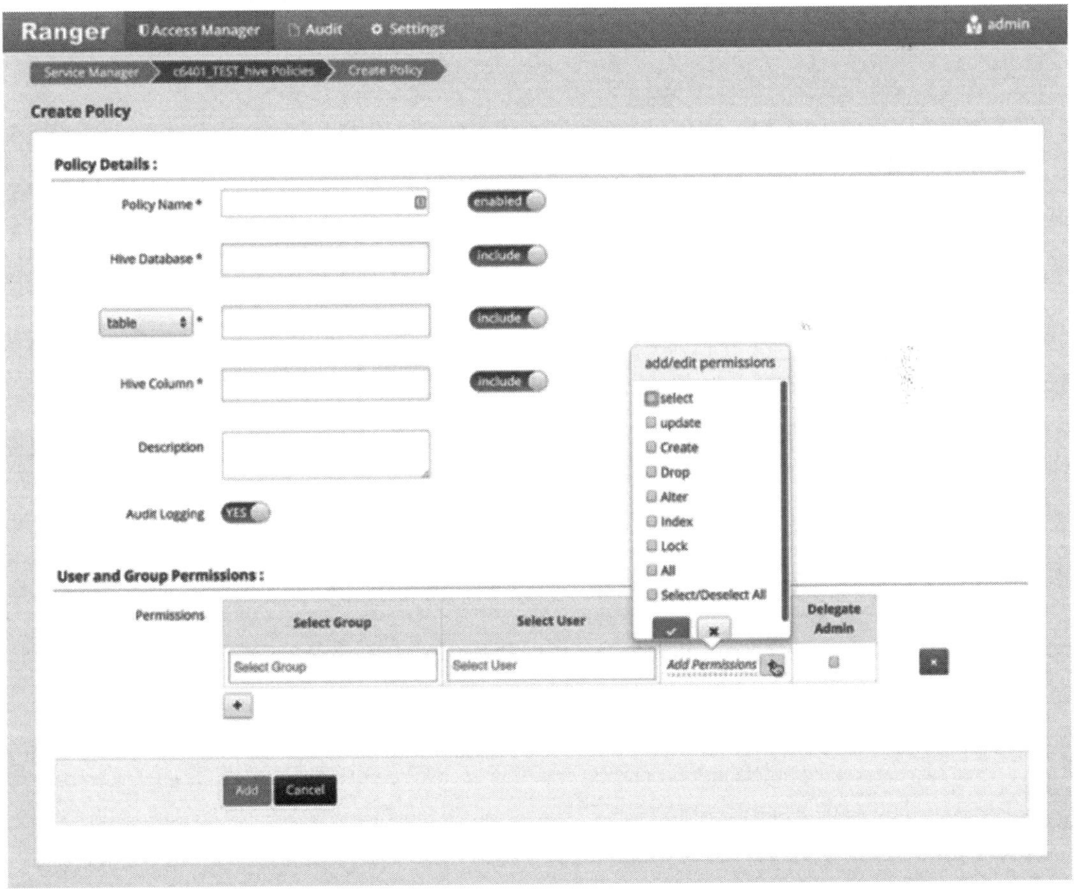

*Figure 10-4. Creating a new Ranger policy*

3.  Provide the following details in the Add New Policy window:

    •   Policy Name—Name that you want to assign to this new policy.

    •   Hive Database—The name of the database for which this policy is defined; you can choose * for all databases.

    •   Table/UDF—Name of the table/UDF; this can be * for all tables/UDFs.

    •   Hive Column—This column is used to control column-level access.

    •   Audit Logging—This parameter is very important as it determines whether the access defined by this policy should be audited or not.

    •   User and Group Permissions—This is where you define the type of access for a user or a group. You can even delegate the admin responsibilities to a user so he can further manage the access of this object.

Once you fill in all the details shown in Figure 10-5 and define the policy, these controls are enforced on the relevant objects in Hive.

***Figure 10-5.*** *Adding details of a new Ranger policy*

■ **Note**    If the Ranger Hive plug-in is enabled and you grant any privileges using the GRANT command in SQL, Ranger automatically creates the Ranger policies for you. This is quite useful when you run a script to create Hive objects and then grant privileges on them.

# Auditing Using Apache Ranger

As previously mentioned, you can also audit various types of access using Apache Ranger. Ranger has a built-in Audit Server that collects all audit data for every plug-in that is deployed. As long as the policy that you created is marked as Audit Enabled, Ranger will audit all access and store its records. These records can then be viewed using the Ranger UI.

In order to see the Ranger audit records, click on the Audit option in the menu bar. You will then see a list of recent audit records, as shown in Figure 10-6.

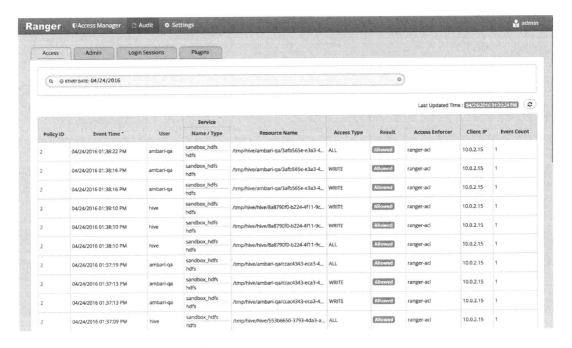

*Figure 10-6. Listing audit records in Ranger*

You can filter these records using various options in the Audit page.

# CHAPTER 11

■ ■ ■

# The Future of Hive

The future of Hive is a roadmap of enhancements and improvements.
The topics of this chapter are:

- Hive LLAP (Live Long and Process)

- Hive-on-Spark

- Hive: ACID and MERGE

- Tunable isolation levels

- OLAP/cube-based analytics

- The HiveServer2 engine

---

■ **Note**   This chapter is a view into the near future of Hive.

---

## LLAP (Live Long and Process)

The demand for sub-second queries calls for fast query execution and lower setup cost of tasks within the ecosystem. The challenge for Hive is to accomplish this without impacting the scale and flexibility that users require from a future distributed solution.

A future-proof methodology using a hybrid engine that leverages Tez and a new engine called LLAP (Live Long and Process) is the next phase for Hive.

LLAP is an enhanced daemon process running on multiple nodes, and it is responsible for the following:

- Caching and data recycle queries with compressed columnar data in-memory (off-heap) copies of the data. Most important speed improvements in the stack to date.

- Multi-threaded execution together with reads with predicate pushdown and hash joins on the Hive ecosystem. Enhances the task allocations and DAG creation.

- High throughput IO using Async IO Elevator with dedicated thread and core per disk to improve the usage of existing environments with more efficient processing solutions.

- Granular column-level security across applications. Hive is going secure without the overhead of other security solutions.

© Scott Shaw, Andreas François Vermeulen, Ankur Gupta, David Kjerrumgaard 2016
S. Shaw et al., *Practical Hive*, DOI 10.1007/978-1-4842-0271-5_11

YARN will be responsible for workload management in LLAP by means of delegation. Queries will transport information from YARN to LLAP about their authorized resource allocation. LLAP processes will then distribute supplementary resources to assist the query as instructed by YARN.

The hybrid engine approach delivers fast response times by efficient in-memory data caching and low-latency processing, delivered by node resident processes. The effective limiting of LLAP usage during the initial phases of query processing means that Hive by-passes limitations around coordination, workload management, and failure isolation that are normally presented by running an entire query in this processing on the databases.

# Hive-onSpark

Apache Spark is rapidly evolving into the programmatic successor to MapReduce for data processing on Apache Hadoop. The successful integration will open the enormous development that is done in the Spark ecosystem directly to Hive.

The biggest is the development in the deep-learning capability of spark. The evolving research into solutions using Spark and TensorFlow will deliver capacity to Hive solutions to use these investments via the Hive-on-Spark stack.

Machine learning has rapidly developed as a critical portion in mining Big Data for actionable insights. Built on top of Spark, MLlib is a scalable machine-learning library that delivers high-quality algorithms.

# Hive: ACID and MERGE

Hive will in the near future support ACID transactions by adding extra functions.
Functions such as:

- INSERT, UPDATE, and DELETE

- Snapshot isolation

- Streaming ingest

Hive will in the near future support MERGE as standard by introducing an Upsert function to Hive. This is a prime improvement to ensure the data warehousing ecosystem is working effectively and efficiently.

The following ACID-supporting functions are coming to native Hive:

- BEGIN TRANSACTION

- COMMIT

- ROLLBACK

Making Hive ACID proof is a massive achievement, as Hive is now successfully hardened for enterprise-level transaction processing.

# Tunable Isolation Levels

A Hive lock manager will be improved to facilitate transactional-level isolation on data transactions. This will empower Hive development to tune the data processing with the best isolation for the specific circumstances.

# ROLAP/Cube-Based Analytics

Analyzing billions of records in near-realtime from within Hive is the future we anticipate. Hive will be empowered with sophisticated, multi-dimensional slicing-and-dicing capability against relational online analytical processing (ROLAP).

This will enable options to construct Kimball bus architectures and the Corporate Information Factory structures within Hive.

Hive will be able to generate SQL and MDX interfaces across the distributed ecosystem with ease to generate cube-based analytics and without negatively impacting the performance of the overall system.

# HiveServer2 Development

Hive clients will interconnect with HiveServer2 over a JDBC/ODBC connection handling multiple user sessions, each with a different thread is the future delivery. Massive improvements in scalability of the core Hive ecosystem are achieved by the new Hive engine.

HiveServer2 will support long-running sessions with asynchronous threads that will speed up the import and movement of data within the Hive cluster.

An embedded metastore for HiveServer2 will performs the following tasks:

- Get statistics and schema from the MetastoreDB.

- Compile queries.

- Generate query execution plans.

- Submit query execution plans.

- Return query results to the client.

# Multiple HiveServer2 Instances for Different Workloads

Hive will in the future be able to adapt to different workloads in a dynamic manner.

Multiple HiveServer2 instances can be used for:

- Load-balancing and high availability using Zookeeper.

- Running multiple applications with different settings.

---

■ **Note**    Hive will evolve into a major component, involved in the building of the future architecture of the distributed data processing ecosystem.

By mastering your processing with Hive skills, you are also securing your future.

---

# APPENDIX A

■ ■ ■

# Building a Big Data Team

Building a Big Data team is a fundamental requirement to ensure the success of business responsibilities for maintenance of production jobs and active projects.

The objectives of this appendix are:

- Describe the basic roles and requirements needed to create a Hive solution.

- Define the minimum group of roles required for an effective team.

- Decide who is assigned responsibility for which element in the solution.

---

It takes time and commitment to achieve a superior solution and get the maximum impact with the team.

---

## Minimum Team

There are a minimum number of roles a successful team needs, as follows.

### Executive Team

The executive team is the main contact between the Big Data team and the rest of the business.

### Senior Sponsorship

The senior sponsor provides the business strategy for completing and maintaining the Big Data program.

They ensure that the business value is achieved through the project work completed by the team(s). They ensure that tasks are value-added and that they enhance the business capacity.

They are accountable for the Big Data solution as a business asset/service to the board of directors. The Big Data solution must be seen as a primary business asset, similar to any other asset listed on the business balance sheet.

### Business Team

The business team is the staff members who form the business support structure for the Big Data program.

© Scott Shaw, Andreas François Vermeulen, Ankur Gupta, David Kjerrumgaard 2016
S. Shaw et al., *Practical Hive*, DOI 10.1007/978-1-4842-0271-5_12

## Big Data Director/Head of Big Data

The Big Data director is responsible for the whole Big Data program to ensure that the strategy from the senior sponsorship is delivered. This person ensures that the entire range of support functions is in place to safeguard future delivery of the services.

The director is responsible for the technical teams that designs, delivers, and deploys the current and future work of the Big Data solution.

## Internal Business Developers and Analysts

The business developers and analysts are the people in the day-to-day business who perform the operational and tactical Big Data work to ensure that daily activities in the business support the longer-term strategy.

These people perform and deliver the solution's business-as-usual activities.

## Technical Team

The technical team is responsible for any technical support for the Big Data solutions. They add new solutions and maintain the existing solution.

## Hive Architects

The Hive architects are the technical owners of the system. They ensure the effective and efficient solution to support the strategy to be designed, developed, and delivered. The Hive architects ensure that complete back-office requirements are provided so the solution is technically sound. They also ensure that future innovative changes do not adversely impact the current solution.

## Hive Administrators

The Hive administrators ensure that the Big Data cluster is performing at an effective and efficient level. They ensure the clusters perform all their technical functions as designed.

## Data Engineers

The data engineers design, develop, and deploy the Hive Extract-Transform-Load process, Reports development, data analysis, and data modeling functionalities.

They assist the architect in implementing the essential modifications to the solution to materialize the strategy of the team into the solution. They are the physical creators of the solution components.

# Expanded Team

As a project grows in size, the team starts expanding to support additional specialists. Specific people are now assigned to specific pieces of the solution. These specialists perform the work necessary to achieve the business strategy.

## Business Team

The business team is the staff members who form the business support structure for the now-expanding Big Data program(s).

## Requirements Specialists/Domain Experts

Experts help the teams from particular business areas ensure the specific areas business requirements are protected by the solution's daily processing.

## Statisticians/Data Scientists

Using advanced data processing methods necessitates specialists in data processing methodologies and statistical analytic solutions.

They ensure that the data processing involves value-added transformations to the business' knowledge using repeatable and verifiable methods.

# Technical Team

The technical team is responsible for technical support of the Big Data solutions. They add new solutions and maintain the existing solution.

## Business Analysts

The bigger team now adds more internal business developers and analysts, but with a more intense role that handles specific business requirement tasks.

The business analyst ensures that the requirements from the specialists/domain experts are accurately recorded and converted into functional and non-functional requirements, which are in turn ready for the development teams to use as guidance.

A large team will use multiple business analysts. We suggest that the project manager organize these specialists into groups of five to eight people with a senior person handling the day-to-day work.

## Data Architect

The data architect is responsible for the data architecture of the analytics systems.

This person uses information technology disciplines for designing, developing, deploying, and managing the analytic data architecture.

Data architects govern in what manner data will be stored, consumed, integrated, and managed by the Hive systems.

There should only be one data architect in an optimal structure. However, for large projects, a maximum of three members can perform this function if they work together as a single design unit.

## Technical Architect

The technical architect is responsible for the server architecture of the analytics systems only.

They use information technology disciplines for designing, developing, deploying, and managing the analytic server architecture as designed by the Hive architects.

There should only be one technical architect in an optimal setting. However, for large projects, a group of five members maximum could perform this function if they work together as a single design unit.

## Hive Developers

The Hive developer is the technical expert who designs, develops, and deploys all the Hive code for the solution. The data engineer develops the data structures into Hive code. The Hive developer optimizes the Hive code specifically for the environment by adding extra optimizations to improve the Hive code.

A large team will use multiple Hive developers. We suggest that the project manager organize these specialists into groups of five to eight people with a senior person handling the day-to-day work.

## Visualization Developers

The visualization developer is the technical expert who designs, develops, and deploys the visualization of the solution.

A large team will use multiple visualization developers. We suggest that the project manager organize these specialists into groups of five to eight people with a senior person handling the day-to-day work.

## Quality Assurance Testers

Quality assurance testers test the system to prevent defects in analytic solution and avoiding defects in the services to users.

A large team will use multiple testers. We suggest that the project manager organize these specialists into groups of five to eight people with a senior person handling the day-to-day work.

## Trainers

The trainer helps the users understand the developed functions of the analytic solution to support the business.

## Technical Writers

A technical writer is a professional writer who writes the technical documentation to help users understand and use the analytic solution.

## Infrastructure Engineers

The infrastructure engineers install, upgrade, and maintain servers.

In large installations, this expert area of responsibility is normally outsourced to a third-party provider.

Cloud services are commonly used in the Hive solution arena, which means that an infrastructure provision could become a simple on-demand request to the cloud provider.

The Big Data director assigns the appropriate responsibility to ensure the Hive solution is covered by a service level agreement.

Remember the team is duty-bound to adapt to the business needs in an effective and efficient manner to deliver value and ensure a successful delivery.

Best of luck with your team's future work on your Hive solution(s).

# Work Lifecycle for the Team

The team should use an agile plan consisting of two sprints of 10 days to add new functionality and then a sprint of 10 days to perform any maintenance releases.

If possible, do not perform new functionality releases of code on the same sprint as a maintenance release. This ensures that the true impact of the maintenance is experienced by the business.

Using the 30-day lifecycle ensures regular delivery of new solutions to the business while supporting a healthy, evolving Hive architecture.

# APPENDIX B

■ ■ ■

# Hive Functions

Hive offers a comprehensive set of functions.
The objectives of this appendix are:

- Highlight the basic Hive functions.

- Explain a simple use of each function.

- Achieve the basic understanding of how to use the functions in a data solution.

## Built-In Functions

We will cover the following classes of functions in this appendix:

- Mathematical

- Collection

- Type-Conversion

- Date

- Conditional

- String

- Miscellaneous

- User-Defined (UDFs)

## Mathematical Functions

| Return Type | Name (Signature) | Description |
|---|---|---|
| double | round(double a) | Returns the rounded BIGINT value of the double. |
| double | round(double a, int d) | Returns the double rounded to d decimal places. |
| bigint | floor(double a) | Returns the maximum BIGINT value that is equal or less than the double. |
| bigint | ceil(double a), ceiling(double a) | Returns the minimum BIGINT value that is equal or greater than the double. |

*(continued)*

© Scott Shaw, Andreas François Vermeulen, Ankur Gupta, David Kjerrumgaard 2016
S. Shaw et al., *Practical Hive*, DOI 10.1007/978-1-4842-0271-5_13

| Return Type | Name (Signature) | Description |
| --- | --- | --- |
| double | rand(), rand(int seed) | Returns a random number (changes from row to row) that is distributed uniformly from 0 to 1. Specifying the seed provides a generated random number sequence that's deterministic. |
| double | exp(double a) | Returns e to power a where e is the base of the natural logarithm. |
| double | ln(double a) | Returns the natural logarithm of the argument. |
| double | log10(double a) | Returns the base-10 logarithm of the argument. |
| double | log2(double a) | Returns the base-2 logarithm of the argument. |
| double | log(double base, double a) | Returns the base base logarithm of the argument. |
| double | pow(double a, double p), power(double a, double p) | Returns a to power of p. |
| double | sqrt(double a) | Returns the square root of a. |
| string | bin(bigint a) | Returns the number in binary format. |
| string | hex(bigint a) hex(string a) | If the argument is an int, hex returns the number as a string in hex format. Otherwise if the number is a string, it converts each character into its hex representation and returns the resulting string. |
| string | unhex(string a) | Inverse of hex. Interprets each pair of characters as a hexadecimal number and converts them to the character represented by the number. |
| string | conv(bigint num, int from_base, int to_base), conv(STRING num, int from_base, int to_base) | Converts a number from a given base to a different base. |
| double | abs(double a) | Returns the absolute value. |
| int double | pmod(int a, int b) pmod(double a, double b) | Returns the positive value of a mod b. |
| double | sin(double a) | Returns the sine of a (a is in radians). |
| double | asin(double a) | Returns the arc sin of x if -1<=a<=1 or null otherwise. |
| double | cos(double a) | Returns the cosine of a (a is in radians). |
| double | acos(double a) | Returns the arc cosine of x if -1<=a<=1 or null otherwise. |
| double | tan(double a) | Returns the tangent of a (a is in radians). |
| double | atan(double a) | Returns the arctangent of a. |
| double | degrees(double a) | Converts value of a from radians to degrees. |
| double | radians(double a) | Converts value of a from degrees to radians. |
| int double | positive(int a), positive(double a) | Returns a for all values of -a and a. |
| int double | negative(int a), negative(double a) | Returns -a for all values of -a and a. |

(*continued*)

| Return Type | Name (Signature) | Description |
|---|---|---|
| float | sign(double a) | Returns the sign of a as 1.0 or -1.0. |
| double | e() | Returns the value of e. |
| double | pi() | Returns the value of pi. |

## Collection Functions

| Return Type | Name (Signature) | Description |
|---|---|---|
| int | size(Map<K.V>) | Returns the number of elements in the map type. |
| int | size(Array<T>) | Returns the number of elements in the array type. |
| array<K> | map_keys(Map<K.V>) | Returns an unordered array containing the keys of the input map. |
| array<V> | map_values(Map<K.V>) | Returns an unordered array containing the values of the input map. |
| boolean | array_contains(Array<T>, value) | Returns TRUE if the array contains value. |
| array<t> | sort_array(Array<T>) | Sorts the input array in ascending order according to the natural ordering of the array elements and returns it. |

## Type-Conversion Functions

| Return Type | Name (Signature) | Description |
|---|---|---|
| binary | binary(string\|binary) | Casts the parameter into a binary. |
| Expected "=" to follow "type" | cast(expr as <type>) | Converts the results of the expression expr to <type>; for example, cast('1' as BIGINT) will convert the string '1' to its integer representation. A null is returned if the conversion does not succeed. |

# Date Functions

| Return Type | Name (Signature) | Description |
| --- | --- | --- |
| string | from_unixtime(bigint unixtime[, string format]) | Converts the number of seconds from the UNIX epoch (1970-01-01 00:00:00 UTC) to a string representing the timestamp of that moment in the current system time zone, in the format of "1970-01-01 00:00:00". |
| bigint | unix_timestamp() | Gets the current timestamp using the default time zone. |
| bigint | unix_timestamp(string date) | Converts the time string in format yyyy-MM-dd HH:mm:ss to the UNIX timestamp and returns 0 if fails: unix_timestamp('2009-03-20 11:30:01') = 1237573801. |
| bigint | unix_timestamp(string date, string pattern) | Converts the time string with given pattern (see here) to UNIX timestamp; returns 0 if fails: unix_timestamp('2009-03-20', 'yyyy-MM-dd') = 1237532400. |
| string | to_date(string timestamp) | Returns the date part of a timestamp string: to_date("1970-01-01 00:00:00") = "1970-01-01". |
| int | year(string date) | Returns the year part of a date or a timestamp string: year("1970-01-01 00:00:00") = 1970, year("1970-01-01") = 1970. |
| int | month(string date) | Returns the month part of a date or a timestamp string: month("1970-11-01 00:00:00") = 11, month("1970-11-01") = 11. |
| int | day(string date) dayofmonth(date) | Returns the day part of a date or a timestamp string: day("1970-11-01 00:00:00") = 1, day("1970-11-01") = 1. |
| int | hour(string date) | Returns the hour of the timestamp: hour('2009-07-30 12:58:59') = 12, hour('12:58:59') = 12. |
| int | minute(string date) | Returns the minute of the timestamp. |
| int | second(string date) | Returns the second of the timestamp. |
| int | weekofyear(string date) | Returns the week number of a timestamp string: weekofyear("1970-11-01 00:00:00") = 44, weekofyear("1970-11-01") = 44. |
| int | datediff(string enddate, string startdate) | Returns the number of days from startdate to enddate: datediff('2009-03-01', '2009-02-27') = 2. |
| string | date_add(string startdate, int days) | Adds a number of days to startdate: date_add('2008-12-31', 1) = '2009-01-01'. |
| string | date_sub(string startdate, int days) | Subtracts a number of days from startdate: date_sub('2008-12-31', 1) = '2008-12-30'. |
| timestamp | from_utc_timestamp(timestamp, string timezone) | Assumes given timestamp is UTC and converts it to given time zone. |
| timestamp | to_utc_timestamp(timestamp, string timezone) | Assumes given timestamp is in given time zone and converts it to UTC. |

# Conditional Functions

| Return Type | Name (Signature) | Description |
| --- | --- | --- |
| T | if(boolean testCondition, T valueTrue, T valueFalseOrNull) | Returns valueTrue when testCondition is true; returns valueFalseOrNull otherwise. |
| T | COALESCE(T v1, T v2, ...) | Returns the first v that is not NULL or NULL if all vs are NULL. |
| T | CASE a WHEN b THEN c [WHEN d THEN e]* [ELSE f] END | When a = b, returns c; when a = d, returns e; otherwise returns f. |
| T | CASE WHEN a THEN b [WHEN c THEN d]* [ELSE e] END | When a = true, returns b; when c = true, returns d; otherwise, returns e. |

# String Functions

| Return Type | Name (Signature) | Description |
| --- | --- | --- |
| int | ascii(string str) | Returns the numeric ASCII value of the first character of str. |
| string | concat(string\|binary A, string\|binary B...) | Returns the string or bytes resulting from concatenating the strings or bytes passed in as parameters in order. For example, concat('foo', 'bar') results in 'foobar'. Note that this function can take any number of input strings. |
| array<struct<string, double>> | context_ngrams (array<array<string>>, array<string>, int K, int pf) | Returns the top-k contextual N-grams from a set of tokenized sentences, given a string of "context". |
| string | concat_ws(string SEP, string A, string B...) | Like concat(), but with custom separator SEP. |
| string | concat_ws(string SEP, array<string>) | Like concat_ws(), but taking an array of strings. |
| int | find_in_set(string str, string strList) | Returns the first occurrence of str in strList where strList is a comma-delimited string. Returns null if either argument is null. Returns 0 if the first argument contains any commas. For example, find_in_set('ab', 'abc,b,ab,c,def') returns 3. |
| string | format_number(number x, int d) | Formats the number X to a format like #,###,###.##, rounded to d decimal places and returns the result as a string. If d is 0, the result has no decimal point or fractional part. |

(*continued*)

257

| Return Type | Name (Signature) | Description |
|---|---|---|
| string | get_json_object(string json_string, string path) | Extracts the JSON object from a JSON string based on the JSON path specified and returns JSON string of the extracted JSON object. It will return null if the input JSON string is invalid. The JSON path can only have the characters [0-9a-z_], i.e., no uppercase or special characters. Also, the keys *cannot* start with numbers. This is due to restrictions on Hive column names. |
| boolean | in_file(string str, string filename) | Returns true if the string str appears as an entire line in the filename. |
| int | instr(string str, string substr) | Returns the position of the first occurrence of substr in str. |
| int | length(string A) | Returns the length of the string. |
| int | locate(string substr, string str[, int pos]) | Returns the position of the first occurrence of substr in str after position pos. |
| string | lower(string A) lcase(string A) | Returns the string resulting from converting all characters of A to lowercase. For example, lower('fOoBaR') results in 'foobar'. |
| string | lpad(string str, int len, string pad) | Returns str, left-padded to a length of len. |
| string | ltrim(string A) | Returns the string resulting from trimming spaces from the beginning(left side) of A. For example, ltrim(' foobar ') results in 'foobar'. |
| array<struct<string, double>> | ngrams(array<array<string>>, int N, int K, int pf) | Returns the top-k N-grams from a set of tokenized sentences, such as those returned by the sentences(). Hive Custom Aggregate Functions (UDAF). |
| string | parse_url(string urlString, string partToExtract [, string keyToExtract]) | Returns the specified part from the URL. Valid values for partToExtract include HOST, PATH, QUERY, REF, PROTOCOL, AUTHORITY, FILE, and USERINFO. For example, parse_url('http://facebook.com/path1/p.php?k1=v1&k2=v2#Ref1', 'HOST') returns 'facebook.com'. Also a value of a particular key in QUERY can be extracted by providing the key as the third argument. For example, parse_url('http://facebook.com/path1/p.php?k1=v1&k2=v2#Ref1', 'QUERY', 'k1') returns 'v1'. |
| string | printf(String format, Obj… args) | Returns the input formatted according to printf-style format strings. |

*(continued)*

| Return Type | Name (Signature) | Description |
|---|---|---|
| string | regexp_extract(string subject, string pattern, int index) | Returns the string extracted using the pattern. For example, regexp_extract('foothebar', 'foo(.*?)(bar)', 2) returns 'bar.' Note that some care is necessary in using predefined character classes: using '\s' as the second argument will match the letter s; 's' is necessary to match whitespace, etc. The index parameter is the Java regex matcher group() method index. |
| string | regexp_replace(string INITIAL_STRING, string PATTERN, string REPLACEMENT) | Returns the string resulting from replacing all substrings in INITIAL_STRING that match the Java regular expression syntax defined in PATTERN with instances of REPLACEMENT. For example, regexp_replace("foobar", "oo|ar", "") returns 'fb'. Note that some care is necessary in using predefined character classes: using '\s' as the second argument will match the letter s; 's' is necessary to match whitespace, etc. |
| string | repeat(string str, int n) | Repeats str n times. |
| string | reverse(string A) | Returns the reversed string. |
| string | rpad(string str, int len, string pad) | Returns str, right-padded to a length of len. |
| string | rtrim(string A) | Returns the string resulting from trimming spaces from the end (right side) of A. For example, rtrim(' foobar ') results in ' foobar'. |
| array<array<string>> | sentences(string str, [string lang], [string locale]) | Tokenizes a string of natural language text into words and sentences, where each sentence is broken at the appropriate sentence boundary and returned as an array of words. The lang and locale are optional arguments. For example, sentences('Hello there! How are you?') returns ( ("Hello", "there"), ("How", "are", "you") ). |
| string | space(int n) | Returns a string of n spaces. |
| array | split(string str, string pat) | Splits str around pat (pat is a regular expression). |
| map<string,string> | str_to_map(text[, delimiter1, delimiter2]) | Splits text into key-value pairs using two delimiters. delimiter1 separates text into K-V pairs, and delimiter2 splits each K-V pair. Default delimiters are , for delimiter1 and = for delimiter2. |

*(continued)*

| Return Type | Name (Signature) | Description |
| --- | --- | --- |
| string | substr(string\|binary A, int start) substring(string\|binary A, int start) | Returns the substring or slice of the byte array of A, starting from start position until the end of string A. For example, substr('foobar', 4) results in 'bar' (see http://dev.mysql.com/doc/refman/5.0/en/string-functions.html#function_substr). |
| string | substr(string\|binary A, int start, int len) substring(string\|binary A, int start, int len) | Returns the substring or slice of the byte array of A, starting from start position with length len. For example, substr('foobar', 4, 1) results in 'b' (see http://dev.mysql.com/doc/refman/5.0/en/string-functions.html#function_substr). |
| string | translate(string input, string from, string to) | Translates the input string by replacing the characters present in the from string with the corresponding characters in the to string. This is similar to the translate function in PostgreSQL. If any of the parameters of this UDF are NULL, the result is NULL as well. |
| string | trim(string A) | Returns the string resulting from trimming spaces from both ends of A. For example, trim(' foobar ') results in 'foobar'. |
| string | upper(string A) ucase (string A) | Returns the string resulting from converting all characters of A to uppercase. For example, upper('fOoBaR') results in 'FOOBAR'. |

## Miscellaneous Functions

| Return Type | Name (Signature) | Description |
| --- | --- | --- |
| int | hash(a1[, a2...]) | Returns a hash value of the arguments. |

## Aggregate Functions

| Return Type | Name (Signature) | Description |
| --- | --- | --- |
| bigint | count(*), count(expr), count(DISTINCT expr [, expr_.]) | count(*) returns the total number of retrieved rows, including rows containing NULL values; count(expr) returns the number of rows for which the supplied expression is non-NULL; count(DISTINCT expr[, expr]) returns the number of rows for which the supplied expression(s) are unique and non-NULL. |
| double | sum(col), sum(DISTINCT col) | Returns the sum of the elements in the group or the sum of the distinct values of the column in the group. |

*(continued)*

| Return Type | Name (Signature) | Description |
|---|---|---|
| double | avg(col),<br>avg(DISTINCT col) | Returns the average of the elements in the group or the average of the distinct values of the column in the group. |
| double | min(col) | Returns the minimum of the column in the group. |
| double | max(col) | Returns the maximum value of the column in the group |
| double | variance(col),<br>var_pop(col) | Returns the variance of a numeric column in the group. |
| double | var_samp(col) | Returns the unbiased sample variance of a numeric column in the group. |
| double | stddev_pop(col) | Returns the standard deviation of a numeric column in the group. |
| double | stddev_samp(col) | Returns the unbiased sample standard deviation of a numeric column in the group. |
| double | covar_pop(col1, col2) | Returns the population covariance of a pair of numeric columns in the group. |
| double | covar_samp(col1,<br>col2) | Returns the sample covariance of a pair of numeric columns in the group. |
| double | corr(col1, col2) | Returns the Pearson coefficient of the correlation of a pair of numeric columns in the group. |
| double | percentile(BIGINT col, p) | Returns the exact pth percentile of a column in the group (does not work with floating point types). p must be between 0 and 1. *Note*: A true percentile can be computed only for integer values. Use PERCENTILE_APPROX if your input is non-integral. |
| array<double> | percentile(BIGINT col, array(p1 [, p2]...)) | Returns the exact percentiles p1, p2, ... of a column in the group (does not work with floating point types). pi must be between 0 and 1. *Note*: A true percentile can be computed only for integer values. Use PERCENTILE_APPROX if your input is non-integral. |
| double | percentile_<br>approx(DOUBLE col,<br>p [, B]) | Returns an approximate pth percentile of a numeric column (including floating point types) in the group. The B parameter controls approximation accuracy at the cost of memory. Higher values yield better approximations, and the default is 10,000. When the number of distinct values in col is smaller than B, this gives an exact percentile value. |
| array<double> | percentile_<br>approx(DOUBLE col,<br>array(p1 [, p2]...)<br>[, B]) | Same as above, but accepts and returns an array of percentile values instead of a single one. |
| array<struct<br>{'x','y'}> | histogram_<br>numeric(col, b) | Computes a histogram of a numeric column in the group using b non-uniformly spaced bins. The output is an array of size b of double-valued (x,y) coordinates that represent the bin centers and heights. |
| array | collect_set(col) | Returns a set of objects with duplicate elements eliminated. |

# User-Defined Functions (UDFs)

```
CREATE FUNCTION [db_name.]function_name AS class_name
  [USING JAR|FILE|ARCHIVE 'file_uri' [, JAR|FILE|ARCHIVE 'file_uri'] ];
```

This statement creates a function by the class_name. JARs, files, and archives will be added to the environment as specified with the USING clause. When the function is referenced for the first time by a Hive session, these resources will be added to the environment as if ADD JAR/FILE had been issued. If Hive is not in local mode, the resource location must be a non-local URI such as an HDFS location.

The function will be added to the specified database, or to the current database at the time that the function was created. The function can be referenced by fully qualifying the function name (db_name. function_name) or can be referenced without qualification if the function is in the current database.

---

Mastering the use of Hive's built-in functions and the permutation chains that you can construct using these functions is of massive significance to becoming skilled at Hive.

These are your data tools.

Practice using them on a methodical basis to grow into an expert at processing data in Hive.

To get an up-to-date reference list, see https://cwiki.apache.org/confluence/display/Hive/ LanguageManual.

---

# Index

© Scott Shaw, Andreas François Vermeulen, Ankur Gupta, David Kjerrumgaard 2016
S. Shaw et al., *Practical Hive*, DOI 10.1007/978-1-4842-0271-5

# Get the eBook for only $5!

Why limit yourself?

Now you can take the weightless companion with you wherever you go and access your content on your PC, phone, tablet, or reader.

Since you've purchased this print book, we're happy to offer you the eBook in all 3 formats for just $5.

Convenient and fully searchable, the PDF version enables you to easily find and copy code—or perform examples by quickly toggling between instructions and applications. The MOBI format is ideal for your Kindle, while the ePUB can be utilized on a variety of mobile devices.

To learn more, go to www.apress.com/companion or contact support@apress.com.

Druck: KN Digital Printforce GmbH · Schockenriedstraße 37 · 70565 Stuttgart